THE TRAVEL IN TIME

SCIENTIFIC VERSION

21 SOLUTIONS FOR 21 QUESTIONS OF THE XXI CENTURY PHYSICS

C. P. FOURNIER

Title:	THE TRAVEL IN TIME
	- 21 Solutions for 21 Questions of the XXI Century Physics - (Scientific Version)
Author:	: Cláudia Penélope Fournier
Publisher / Editor:	Author's edition (C.P.F.)
Graphic Project;	
Cover and Title; Layout:	Penélope Fournier
Translation	Penélope Fournier
Main Cover Image	Oil painting 'Portals'- (author's creation)
Edition:	1st, September 2011
Printing / Distribution:	Amazon.com, Inc.
Legal Copyright Deposit:	122/2009 (Portugal, Lisbon)
ISBN:	978-989-96434-4-4 (paperback)
E-mail:	penelopefournier@gmail.com
Website:	www.cpenelopefournier.com

Chapter 0

PRESENTATION

"Raising new questions, new possibilities, seeing old problems
from a new angle, from a new perspective, requires creative imagination
and brings real progress in Science."
- Albert Einstein -

This compilation is part of a book, a Scientific Romance, entitled: 'The Travel in Time -21 Solutions for 21 Questions of the XXI Century Physics-'. The book portrays the story of a scientist, a researcher of Theoretical Physics, Professor at the most prestigious university in London, the Imperial College. In recent years this teacher has been completely and exclusively dedicated to the compilation of an almost secret project, which consists in formulating a "Final Theory of Time' and with it create a real chance to build a Time Machine and thus achieve the first travel in time made by mankind!

While he walks this path, the theoretical physicist is faced with other puzzles and enigmas of Physics, real problems that no one can solve, problems of Cosmology, Quantum Physics and Relativity, but somehow, this scientist gets the resolution to all these major issues. Seeing in the possession of such revelations the theoretical physicist meets only a few specific members of the university to whom he will propose to explain these questions and reveal this absolutely new theory!

In this small group of elite that he brings together, we can find himself the theoretical physicist, Professor Ruben Klein; a mathematician, Dr. Gibbs; the experimental physicist, Dr. Wolf; a biophysicist, Dr. Stevenson; and an electrical engineer, Dr. Josh Bentley. These person are also fellow teachers and members of the institution and all of these scientists have an important role in the development of the story that can only be revealed at the end...

In the chapters provided in this Scientific Version, was made a selection of the manuscript, a summary of the Theory developed, in order to make it accessible to any reader interested in this topic.

As such, this book can be presented and considered as a work of scientific literature.

In any case, please do forgive me for the bad quality of this translation.

THE TRAVEL IN TIME [Scientific Version]

An Introduction:

The Physics of this century continues to be a science filled with several enigmas. Several questions remain unexplained and many topics are still incomprehensible; fundamental problems are constantly raised and challenging physicists mind. Among puzzles and insoluble experiments, such as the Cavendish experiment and the Double-Fissure experience, fundamental issues are raised, completely defying the understanding, the concepts of logic and human intelligence. Hidden variable persist to stay unknown and can permanently perturb the thought of a scientist.

The 'Travel in Time' has emerged as a need to understand and join together all of the biggest questions currently open in Physics of the XXI century. This topic is extensive by its definition, as such, the main purpose was to present a summarized approach, in a way as complete as possible but rather extensive, of all the major problems affecting the Physics of this millennium. This majestic science is unique and the only one which embraces the atom, the life, the Universe and just about everything, absorbing a vast thematic of topics and different issues apparently divergent and with little relationship. However, relationships, similarities, and truths, are already written in the Big Book of Nature!

This work presents us with a simple Documentary, a Journey and a Travel in Time, simply based on fundamental equations, relations, facts and cosmological considerations, without having to appeal to a too much technical or mathematic language. As such, the entire structure of this book brings together all the essential information and basic concepts of Physics so that any reader can follow what I would like to designate as a Natural History of the Universe!

All solutions presented for the 21 Problems are deeply justified and based with a large scientific support that sustains one single Theory relatively simple…

I think that the 21 solutions could perhaps intrigue the experts of Physics, by raising new questions, doubts and problems and also new ways of solving them.

Basically, this work aims to present a new approach and a new perspective for the treatment and resolution of old problems of Physics.

C. P. Fournier

NATURAL HISTORY OF THE UNIVERSE

All mysteries of Modern Physics revealed

Problems addressed:

This book has emerged over time, with the continued strengthening of knowledge, new ideas, personal research and self-taught education, linked to a deep passion for Physics which has been evolving all alone ... emerging from a free thinking dimension.

" Logic will take you from A to B.
Imagination will take you anywhere."
- Albert Einstein -

Chapter I

INTRODUCTION

" That you may live in an interesting time. "
- Confucius -

We live in a mysterious and fascinating Universe!

While we remain here in this isolated corner of our humble planet, we surrender in to the charms of the Universe and the wisdom of Nature.

And as his faithful disciples that we are, we try at all costs, monitor and understanding the intelligence of the Universe that involves us.

If we were to be observed by Nature with consciousness, what would she say about our progress? ... Are we learning well our lesson?

Indeed a complete theory of Nature requires a great effort on the part of humans ... but people infer and think that the events are not devoid of relation and explanation, so it seems that they are on the right track...

This is the message that we would like Nature to send us!

And so they persist and go on, and still men continue in their incessant search, tireless in their research for a theory for the Cosmos!

The current purpose of science is to create a unity between all particles and forces, including and describing them in a common equation. A Unified Theory would have the ability to give us a full understanding of the physical events that surround us in our Universe, and even of our own existence! A formulation of a Grand Unified Theory lists various properties of reality and Nature, merging them into a single equation. It fields of application would be universal and we will not have to change the theory or the equation to address different problems.

Such an elegant theory would be a genuine work of art!

Only a few great thinkers, guided by a very deep understanding, put the fundamental problem of the structure of the Universe in the center of their thought. And with that in mind, they want to know everything. Those are the real looters of knowledge.

So the ambition persists, and the conquest continues...

Currently, the general description of the Universe is divided into two parts: First, we have the Theory of General Relativity, and on the other hand the Quantum Theory.

The quantum physicists are entirely satisfied with Quantum Mechanics and the astrophysicists are also happy with General Relativity. The Theory of Relativity describes the Universe in an astronomical scale, and quantum theory describes the Universe on a subatomic scale. It is, inexorably, two major intellectual achievements. Both theories are very successful within their area of expertise and each one is supported by an impressive catalog of experimental and observational evidence. However it is known, unfortunately, that these two theories are mutually incompatible. Both theories do not have common solutions that may be applied to the entire Universe!

We know that reality has to be related in some way, so we deduct that there is a missing ingredient, a hidden variable, a property misunderstood, or something that nobody knows. The problem is that everyone knows that we need some changes, but nobody knows how or what kind of change that can make it work.

What I have to present to you here today, is the change that works!

The main science research focuses on the current search for a new theory that integrates both, that joins together Relativity and Quantum Physics … what all the physicists wish to achieve is a Quantum Theory for Gravity!

THE TRAVEL IN TIME [Scientific Version]

There are several models for a theory of Quantum Gravity. Most of them are pure mathematical conjecture without any real physical meaning and purpose, or at least its intelligibility is beyond our reach and understanding. Other, more interesting, use some physical concepts.

Not long ago, in the time of Newton in 1700, it was possible for a person educated assimilate the entire human knowledge ... all human knowledge ... at least in their more general features. Since then, however, the crazy pace of scientific development has made that impossible. Few are the people who can keep pace rising from the frontier of knowledge of the XXI century, always in rapid change and expansion.

What we know now, is probably already outdated and obsolete due to the emergence and flourishing of new discoveries and inventions. As theories are constantly being changed, being updated, acquiring a degree of specificity increased and bigger, all this knowledge and information is the exclusive territory of a specialist and only by him it can be properly assimilated, but even then, they can only hope to assimilate a small part of a model, a very specific and restricted part of a theory with more general concepts.

How can we want to get a proper perspective, the global view, the overall picture, the complete scenario, if we continue to spread the knowledge?

Branches, subdivisions, specifications after specifications, developments that pursue rigorous detail, and for this we stay away from the overall coherence, we delete the relationship between facts, we lose the perspective, we extend our knowledge with new theories, we invent new disciplines ... and we keep subdividing Knowledge ... And, frankly, I must confess ... the Science is exhausted! And full of information!

While this route may seem reasonable and natural, maybe we could include a new branch in this infinite list of new disciplines ... the Interdisciplinary Science!

No one is quite sure why 'c' was chosen as a symbol to denote the speed of light. Perhaps because it is the maximum of the universal constants, or because it has its origin in a Latin word 'Celeritas', which means celerity and rapidity.

But why do our universal constants have the specific value that they have? Why these values and numbers and not others?
Of all the cosmic numbers that define the architecture of our Universe make us realize that a slight change in the value of these constants, however small it was, would have a major impact in the final scenario of our Universe. The final result would no longer be the same and our Universe would be a very different place.

Nobody knows why the fundamental constants of Nature take the numerical values that they assume. These constants are the genes of our Cosmos. This information appears us to be pre-determined from the beginning. These constants seems to have been introduced right from the start, made 'a priori', and it makes us assume an existence of a pre-defined logic that includes and induces a hidden intention to move forward with this magic Universe making it to be the way it is today, and leaves us to believe that there could have exist an 'Architect' for our Universe! Some people think that what makes our Universe the way it is today is related with what they usually call as 'The Anthropic Principle'!

It is surprising that there is no 'Theory of Constants'!

Maybe it was necessary to clarify what is the real role of these constants in the context of Physics, explaining their origins and their impact on the Laws of Nature.

THE TRAVEL IN TIME [Scientific Version]

Probably the only physicist who has spend some time to write a book specifically dedicated to the universal constants have been Gilles Cohen-Tannoudgi. This scientist defends and argued that the fundamental constants, in fact, represent epistemological thresholds, and that their strong connection with the major theories is directly dependent on the proposal for those essential constants as well as the consistency of the choice.

What is the meaning of these constants and what do they reveal exactly? How many are, after all, the real fundamental constants necessary to describe all the Physics?

We say that our Universe is defined by its constants, but we still do not know exactly how many!

We declare that these precious constants are the cause and origin of the interactions of all the fundamental forces that we observe in Nature, but we still do not know specifically which one!

The list of our constants is very long … We could start by mentioning c = speed of light;. then h = Planck constant, then e = electron charge, then perhaps G = Gravitational constant, and why not the most recent discovery α = fine structure constant ... And from here we could continue or we could stop to think and start to ask some serious questions. Why choose a particular set of constants over others?

Currently there are physicists who recognize that the fundamental constants that fall within our Standard Model of Physics are not three, not four, not five, but there are nineteen fundamental constants according to the physicist Michio Kaku, and John Baez more recently estimated that these constants would be twenty-six!

Maybe it would be best if we could review what is our concept of 'universal constant', otherwise, in an increasingly list of new constants always rising, it will set serious restrictions, and it will be very hard to move towards a process of unification and convergence in to a final formula of a Fundamental Theory for the XXI century Physics.

Summarize, simplify, and reduce things to their essence. What we should be searching was for only one constant: The Fundamental Constant of Nature! The magic number that would reveal the secret and the identity of our single Cosmos!

We say that a universal constant is a numeric value, a scalar quantity, which reflects an invariable property of Nature. As such, it is considered as a fundamental essence because it reveals evidence and a guarantee of a particular physical process.

That's what makes these constants unique and strictly universal.

We must never forget that the universal constants are related with physical concepts and not just units. Concepts and different properties of reality are related in such a way that can be merged into one, making disappear a large part of our constants.

Without hesitation, we should proceed and make a good filter, starting with justification, without ambiguities, the link in all our individual constants and their integration and real relationship with a fundamental physical theory.

This selection requires some care and consistency, and great patience, not to proceed and conclude hastily that a particular choice is more important than another. Even more if this options that are chosen because there is not an appropriate theory to match.

On one way or another, we must take risks and make a radical criterion for selection. In the end, we will see that the number of our constants has considerably decreased until a number that is irreducible, making disappear most of the constants and revealing a single and unique constant: The Fundamental Constant!

THE TRAVEL IN TIME [Scientific Version]

Our Fundamental Constant of Nature will have to establish a link between time-space-matter. But unfortunately, the whole construction of Modern Physics is embargoed by a Quantum Theory of Gravity that everyone searches but no one can find!

We exist. Why do we exist?! Was it all a coincidence?

The doubts run over us. Few thinking people will not be asked at some point in their lives if this whole existence could not be a mere fantasy, an illusion!

The more we dissected this world, these atoms, and these particles, the most we found that the apparent strength is a chimera!

The table on which I'm writing is an interlacing of molecules, of atoms ... its constituent particles are part of a core that is 10.000 times smaller than the diameter of the atom! ... ten thousand times smaller.... Among all this space reigns a penetrating Vacuum, permeated only by fields of forces severely strong. What kind of Forces prevails in these fields?

After all, Mass is not the dominant part but the Vacuum is, it is the greatest and the dominant field! And we know so little about the vacuum! This is a case to think! It is amazing how we can think with a brain 'almost' empty!

But, however, things appear to be solid, why is that? What is the secret of matter and Gravity?

To achieve a more definitive conclusion we need to focus on our issue. As such, it is necessary to start with the next question:
What is matter? What is mass? ... And it is around this little question that all takes place...

Chapter II

WHAT IS MASS?

"The poet only wants to stick his head in the heavens.
It is the logical that demands to put the heavens on his head.
And it's his head that gets divided."
- G. K. Chesterson -

The former felt that the basic elements of matter were four: Air, Water, Earth and Fire. The atomists considered that matter was made of indivisible atoms. Today, it is said that matter consists of particles: protons, neutrons and electrons. More recently, the physicists found that particles like protons and neutrons are made of even more elementary particles: the Quarks!

How far reaches the indivisibility of matter? Does Nature have a limit? Or this division will extend forever into the infinity?

The Mass is the key to solve this puzzle, and it is the missing link to establish a relationship between Quantum Physics and Relativity.

To understand the Big Bang, the supposed beginning of the Universe, the physicists have to reconcile the two theories. However, General Relativity and Quantum Mechanics appear to be incompatible. The Theory of General Relativity does not fit in the exact moment of the Big Bang, when our Universe was still a newborn and this time is usually called by the Planck Time; and also does not fit in another time just before a Big Crunch, the physical state that may occur as one of the possible final destinations of our Universe, the death of the Cosmos!

Another inevitable conflict and a particular case in which the General Relativity does not fit is inside a black-hole. In some particular stars the natural evolution consists in an unavoidable gravitational collapse of matter. Within these stars there is a huge density of matter confined in to a very small space. Due to the huge amounts of pressure, density, temperature which is supposed to exist in those physical state, it is expected that the structure of matter as we know, it will not have the slightest chance of survival. And the strong gravitational field will continue forcing a contraction and an inevitable collapse of all the matter and radiation in to the direction of a central point called singularity.

These regions and situations are, in terms of physical behavior, simultaneously big and small, extended and very confined in to an astronomic and microscopic space-time region... and highly complex.

And, indeed, we do not have a good theory that describes what happens inside a black-hole, and also for other physical states with similar conditions, as it is the moment of the Big Bang or the Big Crunch.

And I quote: In these stages "The Gravity is strong. We need a Quantum Theory of Gravity, which does not exist yet. "- Frank Close.
Attempts to combine Quantum Theory with the General Relativity equations transport us in to infinite solutions. If an equation has an infinite solution, the physicists conclude that this has no real meaning in a physical context; therefore, they assume that the equation must have a wrong formulation.

Without a real solution, the physicists have not the slightest chance of knowing what is happening in these physical processes. Using a determinist concept, if we can't understand what happened in the past, we can't predict what will happen in the future.

The Theory of Relativity is no longer valid for the Planck Time and the Quantum Theory also does not provide any solution for these microspaces of high energy. Both theories are in bankruptcy in these circumstances and in this particular regions of space and time.

The existence of a Quantum Theory for Gravity is logical and even necessary!

Problems emerge right from the beginning, as soon as we try to proceed with a quantization of Gravity. This obstacle seems to be insurmountable to us.

If on the one hand the process of physics of particles act on a space-time background, absolute solid and rigid; by the other hand the General Relativity acts on a flexible and dynamic space-time stage.

Considering the Electromagnetic Force, we know that the transmission and mediation of this force is limited to the minimum quantization of photons, and its energy can be summarized to Planck packages. For the Gravitational Force we still don't know what is its mean of transmission, we haven't got the slightest idea how this force takes place, what is its mediator particle, already denominated by Graviton, the mediator of the Gravitational Force, a particle that should be everywhere but hasn't been detected yet; and also we have not found the unit basis, the minimum quantity of matter, the quantization of mass.

This is where we are … at least this is what physicists think.

I can tell you that today we haven't reached much further than Einstein when he aspired for Unification!

The research for a Quantum Theory of Gravity has become a gigantic puzzle and the biggest enigma of contemporary Physics...

The equations are known to each other, some are usually friendly and there are others who disagree violently. When this happens, and two theories persist to get a major confrontation, it is normal that from there arises a third.

The union can be born has a peaceful and unifying concept, the crisis can be an excellence time for creativity and new creation can be born from this state of chaos.

It is often said that a new theory is always an extension of a previous theory ... but it does not need to be necessarily so. When we seek for the truth, we can find many truths in many things!

In the Special Relativity, Gravity was not involved. In order to accomplish a correct theory, Einstein had to include Gravity and take into account all the effects of gravitational force.

Two observational keys led him in to his new Theory of General Relativity:

1 .- The first observation relates gravitational mass of an object with its inertial mass.

2 .- A second observation shows that gravitational field effect can be simulated by an accelerating reference system, even in the absence of gravity.

Now let us describe in detail on what this means:

We say that the gravitational force is proportional to the mass of an object and we say that an object that reacts to the force of gravity is defined as having gravitational mass.

Similarly, a gravitational mass tells us the magnitude of the gravitational force that an object feels.

Furthermore, the inertial mass tells us the speed and how fast an object can move in response to an external force.

Inertia is a measure which reflects the resistance of an object to a change on its state of motion.

For example, an object with twice the mass of another object will feel half the acceleration, when subjected to the same strength and therefore the object of greater mass will move more slowly. In other words, the greater the inertial mass of an object, the slowest it moves when subjected to the same strength.

Take another practical example for this interaction:

If we have a ball of lead and a ball of wood, we say that a ball of lead has a gravitational mass larger than the ball of wood. Therefore, we say that the ball of lead feels a force much stronger when exposed and subject to the force of Gravity, so the magnitude of their weight is bigger.

Furthermore, the ball of lead has a higher inertial mass, this means that it will be slower to react to the external Gravity force.

What this reflects is the following: the force that is applied to the ball of lead may be bigger, but nevertheless the speed of reaction is slower, or inferior, because it has a highest inertial mass; so, therefore, the acceleration is exactly the same as that gained by a wooden ball.

Conclusion: different objects with different masses feel exactly the same gravitational acceleration. And this it is shown in the 2nd Law of Newton:

$$F = ma \Leftrightarrow a = F \, / \, m$$

This ratio (F / m) is always proportional; once again, the acceleration experienced by the two objects is always the same.

The acceleration is the only constant! ... Very interesting!

The equivalence between inertial mass and gravitational mass suggests a deep relationship between these two apparently very different concept from reality. Several experiments were performed and repeated in various scenarios and the results are always the same:

Gravitational mass = Inertial Mass

And this relation can only be explained in one way: if there is no distinction between these two concepts, of course!

Thus, we can establish as definition of Gravitational Mass m_g:

$$m_g = F / a$$

And, besides, we can give another definition of inertial mass:

$$m_i = F / a$$

At first glance, there is no obvious reason for these two types of mass to be related. The gravitational mass is the capacity that an object has got to attract other, and is usually expressed in Newton's Gravity equation ($F_g = G.m^2/r^2$), relating the magnitude of gravitational force. The inertial mass is the one expressed in the second Newton's law of dynamics ($F=ma$), which relates the movement and speed.

But the facts do not lie and experience proves that these two separate measures are usually confused and merge it into one.

The conclusion can only be one: that these concepts should be identical in their essence and therefore they can be interchangeable.

Now for the 2nd point:

23

To clarify a little better the relationship between gravitational field and acceleration of a reference system, we will show a practical and concrete example:

When we are on board of a plane, just about to take off, we all can feel our own weight that push us down, but also an additional inertial force that makes us feel even heavier. This force arises when the plane starts to accelerate, reaching enough speed, the sufficient velocity to lift off. When the plane takes a constant speed, a cruise velocity, everything returns to normal and we feel only our own weight.

The opposite effect can also occur: If the plane takes a free fall, in this new and accelerated reference system, we are without weight and we no longer feel the effects of gravitational force.

With this experience we can also suggest that there is a deep connection between Gravity and accelerated reference systems, such that:

Gravity = acceleration

Also these two concepts must have some complicity in its essence and can, therefore, be interchangeable.

For those who have never traveled by plane, another similar example occurs on board of an elevator. In the particular case of free fall, we do not feel our own weight because we fall with the same acceleration as the elevator.

Based on this type of observations Einstein concluded his Principle of Equivalence, summarizing:

The definition between the masses of two objects is defined in Mechanics of two different ways: by the inverse reason of the accelerations that are communicated by a same force (Inertial Mass); and also by the direct reason of different forces that are applied in two distinct objects when exposed into the same gravitational field (Gravitational Mass).

The equality and similarity of these two kinds of mass, identified by a very different method has not raised any issues or questions for the current physicist. However, Classical Mechanics does not give any explanation about this type of equality!

Is it simply a gap?

Is there a possibility that these equalities are trying to reflect the true nature of these concepts?

A little reflection shows that this Principle of Equivalence is extended to the Principle of Relativity, which means that can be applied in to coordinated systems with non-uniform motion, accelerated systems, with relative movements in relation to each other.

Let us see how:

If we consider a system of inertia K, where all objects are sufficiently distant from each other, we could say that, on this system K, there isn't any kind of acceleration, so everything remains at rest and all the particles remain at the same position.

But if we consider another reference system K ', uniformly accelerated, moving with non-constant speed; from this point of view we could say that all the masses of reference K have all equal and parallel acceleration; we will consider that those particles are all moving away from the reference K ', and behave as if they were subject of a gravitational field! And yet, as if K' did not have the initial acceleration considered!

In other words, assume or accept that K' is at rest and that in that region there is only a gravitational field, is the same as believing that K is the legitimate reference and that there is no gravitational field in K', and that this one is only accelerated!

What this example means is that nothing can be concluded, therefore, in space there is no fixed reference point, there is not an absolute reference system for which we can make measurements and establish absolute physical facts.

The Principle of Equivalence proposed by Einstein states that the laws of Physics in a Gravitational Field are exactly the same as in an Accelerated Reference System; establishing that we cannot make any distinction between both. In fact, we cannot do any kind of experience that confirms us, or let us know, in which situation we are.

This immediately raises another question about the Universe:

After all, where do we stand?

Immersed in a huge gravitational field, or transported in a huge accelerated system? Since we can considered Gravity as a Natural Inertia and Inertia can also be considered like an Artificial Gravity.

Interesting idea indeed!

Another link to be taken into account as we try to reveal the mystery of matter, is the following:

For a long time scientists have considered Mass and Energy as two different phenomenon. What science has taught us is that mass and energy are indestructible and that they both satisfy identical Laws of Conservation.

Einstein, once again, had the vision to see that both mass and energy had exactly the same characteristics, curious as ever, noticed that both are contracted and expanded in similar factors, their properties were very similar. In all major aspects he concluded that Mass and Energy were indistinguishable, that inertial mass is simply latent energy. And with a final revelation he has shown us that mass could be destroyed and converted into energy and vice versa.

And present us the Mass-Energy concept through his famous formula that we all already know:

$$E = m.c^2$$

The experimental and incontestable evidence of the demonstration of this formula is in the disintegration of radioactive elements and in the nuclear fusion of stars.

But it should be noted that Einstein incorporated these two concepts in only one, introducing a new Principle of Conservation of Mass-Energy, or more simply, the Law of Conservation of Energy.

From a practical point of view, about the Principle of Equivalence between mass and energy, this means that a material object can be transformed into pure motion; Kinetic Energy can be converted into mass and vice versa. What this means is that pure movement can be transformed into a pure solid object!

If you notice well, this is truly amazing!

Matter can be created only by spending motion … movement … energy? … Very interesting!

This transformation would only have to obey to the Law of Conservation of Energy.

We say that Energy = Mass x Speed of Light squared, meaning that matter can be destroyed but it can also be created and transforming the equation, we have:

$$m = E / c^2$$

This is not a new formula, but with this equation we can see how false it was the belief of the past that considers that matter could not be created nor destroyed, and how false is the belief of today about the stability and quantization of matter, because this physical concept it is not even a fundamental property, but only apparent!

Indeed, there is no Law of Conservation of Matter, what is always preserved, as far as we know, is the Energy!

The only real quantities that are always conserved in collisions are the momentum (quantity of movement) and energy, which means, the theoretical conservation of mass.

However, we aspire for the quantization of matter, but then we make a relationship between matter and energy; we admit that these two concepts are indistinguishable and the same. And now we separate the same concepts and relate Mass with Gravity and Energy with Field. So, after all, where are the differences and the similarities? I do not understand … is it possible that we are those who are imposing the differences?

We could begin by clearly conclude that the mass of an object is a measure of its energy content. That the Mass is a form of Energy! That Mass is a manifestation of Energy in Motion!

That would simplify many things! And this is what Einstein tells us, when we read his equation.

Energy together with its component vector: momentum or amount of movement. Energy in motion, these two concepts go hand in hand together and both produce everything that exists! Everything is energy in motion...

Before finishing, let's see another analogy: If on one side, the inertia of an object depends on its energy content, such that:

$$m = E \, / \, c^2$$

And on the other hand, with a little help of quantum theory, we know that E = h.f, so the energy depends on frequency, and replacing our equation of inertial mass, is that:

$$m = h.f \: / \: c^2$$

Moreover, it's also known that there is a relationship between mass and momentum (p=m.v), so that the equation of classical physics that relates these two quantities is:

$$m = p \: / \: v$$

This equation says that the mass of an object is a measure of its speed and its momentum.

And finally, by the famous Newton's formula (F = m.a), it is known that the mass of an object is a measure of its acceleration, whose origin is in a mysterious force, supposedly Gravitational:

$$m = F \: / \: a$$

I do not wish to add anything else to the concept of mass, because if we do so, we will start to get confused. Indeed, everything is already said: Basically what we have seen until now is that Mass can be described by several different equations, relating different concepts and different units of Physics, and that the matter may appear in a very diverse and versatile way!

Unless the mass is not an innate characteristic, a fundamental unit, and essential concept; it follows very clearly that this mysterious reality and fact of Nature cannot be explained and described in an objective way! Therefore, trying to summarize ... all these versions of mass lead me into a question: Does mass have some kind of a multiple-personality syndrome or disorder?!

Or is that all these relations and equations translate and hide its true physical meaning? ... Before we move forward to any kind of conclusion in our relation and definition of mass, perhaps we could ask for some help to Gravity...

Chapter III

WHAT IS GRAVITY?

"Learning is the only thing that mind never gets tired,
never is afraid and it's never sorry."
- Leonardo da Vinci -

The 'gravity' of the situation, is that this Gravity Force remains very difficult to explain, since it does not fit with most any known theory.

The gravitational force is very unique, different and original and apparently independent of all the other forces.

It is because of this force that was born the concept and status of mass and, consecutively, the whole structure of matter, of atoms, molecules ... and Life!

What is the mystery of the Force of Gravity? ... After all, it exists!

Again, let us start from the beginning...

The Newton's Mechanics was so successful in the XVIII, XIX, XX and XXI century, which virtually ceased to be questioned. With this theory we predicted solar eclipses, sent men to the moon, and put space probes and artificial satellites operating in the Universe!

The perception of Newton took into account the Earth-Moon system and the fall of an apple from the top of a tree, linking them, he concluded that the force of attraction that makes the Moon remains in its path is the same that makes the fall of the apple.

He wrote then:

"I have concluded that the forces that keep the planets in their orbits are in the ratio of the reciprocal squares of distances to the centers around which they are orbiting and, therefore, I have compared the necessary force to keep the moon in its orbit with the force of Gravity on the surface of the Earth and found that the two solutions are almost equal. "- Sir. Isaac Newton -.

An episode that led Newton to imagine that, perhaps, all the objects of the Universe felt the same force of attraction and, as such, they were all attracted by each other at the same way, and with that in mind he wrote the first Law of Gravity.

A bright idea and a real revelation, no doubt ... for the epoch and the time in question, however, today, we haven't progressed much more!

Until now, we have failed to reconcile the theory of gravity with quantum physics, because a simple reflexion shows that the basis of the formula of Gravity is incorrect and the principles upon which they are based have several fails, right from the start!

It is said that Gravity is the force responsible for the cohesion of subatomic particles, for the attraction of mass.

Taking up the steps of Newton, we would say that Gravity is a force that acts in a way inversely proportional to the square of the distance, depending on the relationship $1/r^2$. This means that if we take, for example, the distance of 1m the gravitational force has a specific value, replacing r by 1 we have $1 / 1^2 = 1$ and at 2 meters away from the gravitational force is $1 / 2^2 = 1 / 4$, which means that this force will be four times smaller and so on. And we could conclude that the effect of gravitational force is lost and attenuates with the distance and separation from the source. Right?

So, again, we can conclude that the more distant from the Mass, the weakest is the force. Similarly, we could suppose the inverse way: the closest to the center of gravity of any mass the strongest is the force.

Continuing with this logic, this would imply that in the center of any object, the gravitational force would tend into their maximum force, into a very high value, an infinite value!

If these situations occur in practice, the gravitational force in the center of any object would be very strong, infinite, and everything would collapse into black-holes and the Universe could not even exist!

The current physicists are perfectly aware of this issue, but they don't seem to give much thought in to this particular problem in the law of gravitation! They simply ignore it!

The Theory of Gravity is not valid for r = 0, because in this case we have:

$$F_g = \infty$$

How come that a fundamental formula that establishes relationships between mass and matter, cannot establish the existence of matter itself?!

It seems to me that there is a great inconsistency in this Theory!

The physicists claim that this particular formula is not valid for specific situations called singularities, such as the Big Bang, Big Crunch or black-holes. But what we can see is, in fact, that this it is not a valid formula for the Universe that surrounds us every day, just like the way it is!

Assuming that we could conceive a mutual attraction to essential matter, then, in a primary state of the Universe, we would assume that this Gravity Force would be uniformly distributed, and the primary attraction would reach all of the primitive particles in the same way.

With a model that considers that the Universe began with an innate Gravity force, the most fundamental particles could never have a proper

33

evolution with an independent concentration of mass. The implementation of a single individual particle, the execution of an independent atomic structure, the generation of protons, neutrons and electrons would be nearly impossible. All of this because the primordial density of matter would also be evenly well distributed and the innate Gravity field would attract all of these particles uniformly towards to a central point. The whole mass would be concentrated into a single structure and the whole matter would all be placed in a central and densely compact object, a singularity ... and that would be the only object existing in this Universe!

As such, we find that matter could never emerge and evolve under these conditions, because the inherent Gravity of the Universe would not allow the existence of a single individual and consistent atomic structure!

But the Universe exists, it is composed of galaxies, planets and solar systems ... and it is minimally stable, so there must be some explanation for this has not happened!

Neither the density fluctuations can explain the process of evolution which formed bigger structures of matter. The fluctuations are slow, very slow, the power of the gravitational attraction is almost instantaneous.

The answer to this puzzle relies in the definition of mass. Only that can explain the great enigma of the origin of matter...

Chapter IV

THE ORIGIN OF MATTER

*"There are more mysteries between heaven and earth
then those which dreams our vain philosophy. "*
- Shakespeare -

The matter is a very exotic property of the Cosmos.

It seems to me that the best and most logical definition of mass will come from Einstein's equation $m = E/c^2$. We could see Mass as a kind of Energy materialization! This expression would be responsible for the formation of particles of radiation and particles of matter.

My suggestion goes in the direction of proposing that the materialization of the energy has been obtained from a pure energy that already existed in the early stage of the Cosmos. And the consolidation of the first material-energy particles would have evolved from a state of chaos until it reaches a minimal state of balance and order. Finally, the consolidation of mass is nothing less but pure energy in motion.

Before we proceed with the explanation of this process let us look at the following experience where a similar effect occurs. This experience shows that a natural evolution of a system that starts with a state of chaos but, with time, the system reaches a state of natural equilibrium, balance and order. For example, when considering the heating of a liquid, we can see, initially, a chaotic thermal disorder, in which the molecules travel in all directions in a disorderly manner. But if we continue to heat this liquid more and more, then, from a certain moment, we can observe the formation of small vortices of constant rotation, in which billions of billions of molecules follow each other in to a perfectly orderly movement.

This experience can be verified experimentally and it notes that there is a natural formation of order from an initial chaos condition in a spontaneous way. Nature follows its natural meaning, which is to achieve the minimum state of balance and equilibrium, thus, by creating these constant movements of rotation Nature avoid a state of absolute chaos and maximum entropy.

With this analysis we will assume that a similar process occur in the formation of the first particle of matter in our Universe. Suggesting that, primitive particles have emerged, not from a liquid but from another specific substance. This substance would be a primitive kind of energy already existed in the beginning of the Universe.

It would be proper to remember that by Classical Physics, we say that there is an electromagnetic or gravitational field in a specific region of space when an electrical charge or a mass is placed in that region and exposed to these fields and as a result it feels the effects of these forces.

With this idea we can highlight the following conclusion, which is: even in the absence of charges or masses, at least there are fields!

Our Primordial Universe would be emerged in a primitive field. And even in the absence of electrical charges, masses, quarks or any kind of particles such as photons, at least there would remain a Pure Primordial source of Energy and Radiation.

The lines of force of a field are areas that represent a virtual storage of energy, lots of energy, potential energy. The lines of energy are a Field of Force, which is also a Potential Field, which means a virtual storage of an invisible energy that can only manifests itself when masses or charges are introduced. But this field is always present, potentially active, waiting, patiently, for an opportunity to become real and manifest itself. This is the mystery of a Field Force!

Resuming, we could say that the kind of force which was dominant in the beginning of the formation of the Universe is related with Pure Radiation or Pure Filed Energy. This would be the substance that existed at the beginning of the Cosmos ... Pure Energy in Motion, that is, momentum. Considering that:

$$E = m.c^2$$

$$\Leftrightarrow m = E/c^2$$

And furthermore:

$$p = m.v$$

$$\Leftrightarrow m = p\,/\,v$$

Matching both equations, it emerges a new relation which is:

$$E/c^2 = p\,/\,v$$

Assuming that in this period of time the Universe would be immersed by high amounts of energy, filled with pure primitive radiation, in these conditions these virtual particles would reflect a high quantity of kinetic energy and most of its energy would be closed in its frenetic movement, constantly shaking from one side to another. Replacing in the equation v by c, is that:

$$p = E.c / c^2 \Leftrightarrow p = E / c$$

Again, momentum 'p', amount of movement, which is no more than Energy in Motion. This was the initial state of our Universe … pure energy in motion…

Momentum and Energy! The only two units of Physics truly fundamental…

In a continuous and random movement under these high densities of energy, some of these movements would develop a Rotation Movement. And this rotation is extremely important. These little vortices of rotation may be associated with the first quality inherent to any particle which is a very important property of matter: the Spin!

All known particles by the physicists have three properties: the mass, the charge, and Spin. We can imagine the spin as a characteristic associated with the movement of energy rotation and with the magnetic moment of these particles. In a way, we can define spin as a direction in space, spin defines a direction of an axis … but this characteristic that all the fundamental particles present is truly unique and hard to explain … All particles have a spin very well defined, and this quality does not change, nor can it be changed, it is permanent and immortal.

Let us now present another analogy: We could remember that the process of magnetization of a cylinder of steel can be obtained in the absence of magnetic fields (except the field produced by the Earth), making the cylinder spin rapidly on its axis. As faster as it moves, as stronger is the magnetization process. At the end of this continuous movement we will find a new magnetic field concentrated on the surface of the cylinder. This phenomenon is in agreement with the

Electromagnetic Theory, in which any electric charge in rotation generates a magnetic field in its surrounding. A macro example of this effect is the terrestrial magnetic field, which origin comes from the movement of charges within the nucleus of the Earth.

In our particular case there is no electrical charge, only movement and energy. This continuous movement of energy in rotation takes a specific and constant acceleration, the same as in the cylinder it forms a spherical field above the surface, surrounding a center point. We could designate this field as a kind of Surface Gravity Field, below this surface there is no gravitational field, which means, where the radius = zero, Gravity is also zero, and there is also no Gravity forces acting in any part closest to the center.

Thus the idea of this model solves the problem of the infinite in the formula of Newton, stating that Gravity does not exist inside the particles, assuming that this is a Surface Field and external to the particle.

This would be the process of formation of all fundamental particles!

This concept would end the infinite indivisibility of matter!

This surface field can be consider as a shield, an armour, a blindage which is much stronger in the border area, as if forming a wall of strength, an extremely strong field, but in reality is invisible and immaterial.

The particles delude ourselves ... make us believe that the matter is material that it has got mass in its constitution ... all illusions!

However, by this stage I wouldn't call it as a Gravitational Force, it would be more appropriate to designate it like a Material Force, responsible for the formation of matter. Since this material force does not have any attraction characteristic, it is simply a blindage, and it has got no charge, or else, to be more precisely, it has got a neutral charge.

This material structure consisting of small Force Vortices of high density energy, would fill several points of space in the Primordial Universe, when our Cosmos had only a few microseconds of existence.

We could assume that these first primitive particles would be a kind of primitive rotators (neutral particle of rotation), but relatively dense and particularly unstable and vulnerable. In a constant frenzy of fast movements and high energy, these primitive and neutral rotators would be broken. From a result of these shocks and collisions it emerges some new particles, more stable, specific particles that are called today like the fundamental particles of matter: the Quarks.

The Quarks are, most likely, the more stable and the most fundamental particles of the Universe.

Following the patterns of a Natural Process of Evolution, towards the survival of the strongest ones, our primitive rotators would have a very short life time period and a temporary survival. Unstable, these primitive rotators would continuously disintegrate in a process of transformation in order to achieve a more stable particle: a Quark.

This spontaneous breaking has its origin in a very fundamental force, the force of instability, which is simultaneously, a Force of Balance. This is an extremely Weak Force, but with a critical and fundamental part in the evolution of the Universe.

This is the very first sign of the interaction of the Weak Force, which would lead to all neutral rotators to disintegrated and disappeared, leaving behind: a sea of Quarks.

In this Universe, more expanded, the temperatures began to fall and this new environment has brought some peace and serenity to these particles. This shiny glimpse of stability allowed creating order from chaos.

According to the standard cosmology, the physicist state that:

"A Universe with only three minutes old had formed its first atomic nuclei."

From my point of view, I must say that this seems a little bit early. The creation of a single atom involves a great level of complexity.

It is known that the constituents of the nuclei are protons and neutrons, and that the constituents of these particles are the quarks, or rather, triplet of quarks. Each proton is formed by a group of three quarks, and each neutron is formed by a group of three quarks. All quarks have the same spin = ½. But quarks are not all the same!

First, they are extremely small, the order of 10^{-18} m, therefore, very difficult to detect by direct experience. Indeed, they are so small that its size is not yet well established, and there is no direct evidence of this substructure, we can only infer its existence because they fit perfectly on a theoretical model of the structure of matter.

Another characteristic that arises is: all quarks have electric charge. And this is another very fundamental characteristic of matter.

Interestingly, the quarks must be the first particles responsible to cause the formation of the electrical charge. The rupture of the primitive particles of rotation, may have led to the rupture of the charge itself, dividing it into fractional charges.

The truth is that experiments have proved that a neutron consists of three quarks, with two charges equal to -1/3 and one charge equal to +2/3, which makes a total amount of a neutral charge, which is the charge of the neutron. Adding everything, the final charge is equal to zero …

$(-1/3) + (-1/3) + 2/3 = 0$

Similarly, the proton consists of three quarks, two charges equal to +2/3 and one charge equal to -1/3, which makes a positive charge 1, which is the charge of the proton.

There are only two types of charge in quarks: $+2/3e$ and $-1/3e$.

The first unification of Nature must have emerged towards the quarks unification producing the union of these fundamental particles. The first agglomeration must have followed to recover the balance, which is to achieve the stability of the electric field. As such, the favorite group would be those build of three quarks that could form a stable and electrically neutral particle: the neutron.

This small electrical attraction led, most probably, to the formation of another type of group or groups of quarks, also neutral, but more instable, build with a higher number of quarks, but that, somehow, this specific group did not satisfy the requirements and interests of Nature.

This primitive form of matter would be unsuccessful in the path of evolution and it would not have evolved, remaining hidden and obscure, distributed by a vast area of the Universe, it would be some kind of a failed matter ... really interesting ... matter hidden, matter black, dark matter...

Nature is smart. Do not underestimate the intelligence of Nature.

The Evolution follows always its way to acquire and achieve a higher degree of complexity. Everything has got a reason to be. Remember that, by the concepts of Darwin, the natural path of primate's evolution ended into different ways; emerged in two distinct branches: Chimpanzees and Humans.

There is much missing matter in the book of accounts of the Universe. Recently physicists have discovered a mysterious type of matter, so far unknown, because its presence has remained hidden during all this time ... it is the exotic Dark Matter!

Chapter V

WHAT IS THE DARK MATTER

"Discovery consists of seeing what everybody has seen
and thinking what nobody has ever thought. "
- A. Von Szent-Györgyi -

Is it possible that we can find out what is this exotic matter?!

Scientists estimate that only in the next ten years we will build appropriate equipment and instruments capable of isolating the dark matter and reveal this great mystery of our Universe.

Looking for our planetary system we can deduce the speed of the planets around the Sun, considering only the mass and gravitational influence of the star king, the Sun. According to the laws of Kepler, so the planets can maintain its stable orbit, they must acquire a certain speed, and this period of translation has different values depending if the planet is closer or more distant from the Sun. So, Mercury has a very rapid movement of translation, while Pluto, a more distant planet, has a rather slow movement of translation. This means, in practice, that we can deduce which is the peripheral speed of a system due to the gravitational influence of the central star.

And this was exactly what Fritz Zwicky tried to calculate. When measuring the peripheral velocities and distances from the galaxies center of hundreds of thousands of galaxies like the Milky Way, he found out surprising results: he has discover that the mass of the system was 100 times higher than the one estimated based only on light emission from galaxy!

If you counted the amount of visible matter in a galaxy based only on the emission of light sources, the calculations showed a much lower value than that obtained experimentally and confirmed when measuring the peripheral speed of the galactic system. This is definitely a very surprising result!

It appears to be that there is a mysterious kind of matter, which we can't see, surrounding the galaxies and creating a lot of Gravity.

This strange matter, Dark Matter, which does not emit light, but more surprising is that it does not emit any kind of radiation! Neither Infrared nor Ultraviolet, even X-ray or Gamma radiation ... or anything! Any atom known today processes some kind of radiation!

And now we ask, how can we look at something that we can't even see? ... The answer to that is very simple ... by the gravitational effect that produces, which affects directly any massive system.

This strange form of matter seems to have the capability to produce, indeed, immense Gravity. Currently, all astronomers agree and concluded that 90%, or even more, of all mass of the Universe which is capable to exert gravitational forces does not emit any trace of light. Ninety percent of matter in the Universe consists of this very mysterious Dark Matter! It is a lot of matter...

The Universe is dominated by a totally unknown and mysterious kind of matter, the acclaimed Dark Matter. Nobody knows what this substance is. Where does this matter come from? And more strangely is that it does not emit any electromagnetic radiation!?

Scientists dismiss the hypothesis of this strange substance to be some kind of atom or chemical element and they concentrate on particles. They already have designated enough names for this dark substance, since particles called WIMP (Weakly Interacting Massive Particles), and other candidates such as Neutralinos and Axions.

No one is quite sure whether this problem belongs to the field of Particle Physics or Cosmology. I would say that the solution to this mystery relies in the hands of Chemistry!

In order to reveal this mystery, it would be enough and sufficient if we concentrate on its most peculiar characteristic: the absence of radiation. … I will leave you to reflect about this for a moment …

Before we continue, it is necessary for us to bring in mind the following table:

0⁻	1	2	3	4	5	6	7	8	9	10	11	12	13	14	15	16	17	18	0⁺
A																			Bm
	1 H																		2 He
	3 Li	4 Be											5 B	6 C	7 N	8 O	9 F	10 Ne	
	11 Na	12 Mg											13 Al	14 Si	15 P	16 S	17 Cl	18 Ar	
	19 K	20 Ca	21 Sc	22 Ti	23 V	24 Cr	25 Mn	26 Fe	27 Co	28 Ni	29 Cu	30 Zn	31 Ga	32 Ge	33 As	34 Se	35 Br	36 Kr	
	37 Rb	38 Sr	39 Y	40 Zr	41 Nb	42 Mo	43 Tc	44 Ru	45 Rh	46 Pd	47 Ag	48 Cd	49 In	50 Sn	51 Sb	52 Te	53 I	54 Xe	
	55 Cs	56 Ba	57-71 La*	72 Hf	73 Ta	74 W	75 Re	76 Os	77 Ir	78 Pt	79 Au	80 Hg	81 Tl	82 Pb	83 Bi	84 Po	85 At	86 Rn	
	87 Fr	88 Ra	89-103 Ac**	104 Unq	105 Unp	106 Unh	107 Uns	108 Uno	109 Une	110 Unn	111 Uuu	112 Uub		114 Uuq					

La* Rare earth metals (Lanthanide series)	57 La	58 Ce	59 Pr	60 Nd	61 Pm	62 Sm	63 Eu	64 Gd	65 Tb	66 Dy	67 Ho	68 Er	69 Tm	70 Yb	71 Lu
Ac** (Actinide series)	89 Ac	90 Th	91 Pa	92 U	93 Np	94 Pu	95 Am	96 Cm	97 Bk	98 Cf	99 Es	100 Fm	101 Md	102 No	103 Lr

- Periodic Table of chemical elements -

Here it is! The Periodic Table of Mendeleev! A true masterpiece of its author!

When Mendeleev discovered that all the elements of Nature follow a pattern, he must have been ecstatic and delighted! And all of this with a simple pack of cards...

As you can see Nature is organized. The chemical elements are all divided according to their atomic number. This number represents the number of protons in an atom, and this is a quality which characterizes and distinguishes all elements of Nature.

So that, if a particle is composed of eight protons, eight neutrons and eight electrons, we know that we are talking about the Oxygen, whose atomic number is 8.

Similarly, if we have just a single proton, one neutron and one electron, we know that we are talking about Hydrogen, whose atomic number is 1. This is the simplest element of Nature. Correct?!

When we refer to chemical elements we can also include another feature that is common to all of them: the number of neutrons.

A stable atom has, always, the same number of protons, neutrons and electrons. Except in individual cases in which there is an unbalanced number of neutrons, and then we designate these atoms of isotopes, or when there is an unbalanced of electrons, and there for these atoms are called ions.

The secret balance remains in this perfect triangle:

number neutrons = number protons = number electrons

Right?

Let us remember a little better the composition of our Periodic Table. Indeed, there is one chemical element that avoids this pattern!!

Most curiously the Hydrogen does not share this sacred triangle! ...I wonder why that is! ...

Why is that? The most common and simplest element of the Periodic Table, Hydrogen, it is, strangely that it might seem, an Isotope!? This means that the Hydrogen that we usually find in Nature has lost a neutron!? What happened to this neutron? This is a fact that has been ignored ... is this an accidental coincidence of Nature? Nature does not have many coincidences without a reason...

Track number 1: What is that Dark Matter and Hydrogen Isotopes do have in common?

I'll leave you to a moment of reflection...

And the answer is ... nothing at all! And that is exactly it!! The Dark matter could not proceed in the development of a natural path of evolution. This means that these primitive neutrons were never able to process protons; did not share the capability of a neutron/proton mutation process. Therefore, The Dark Matter has got an atomic number zero because these primitive neutrons never managed to create protons. The particles of Dark Matter will be our 'Chimpanzees' and the Chemical Elements will be our 'Humans '.

The concept of Darwin's evolution does not only apply to biological organisms. The process of mutation allowed the neutron-proton evolution.

Let us now have a more abstract moment ... traveling through time ... a travel in time to try to look and see what is happening in this period of our Universe.

We enter in a dark Universe, very dark ... and virtually immobile, static and homogenous ... but full of energy and with a lot of potential!

There is a mysterious force that is already present in this almost ghostly Universe, and that it is taking shape, this is: the Weak Force, the subtle Force of Instability and simultaneously the Force of Balance … the force responsible for changing neutron in to a proton.

As we know, one of the forces present in Nature is the Weak Force. We know very little information about this force. But with what we know, we deduce that this is the force responsible for the disintegration of radioactive elements.

We know that some chemical elements are radioactive because they are unstable, or rather, the instability produces radiation!

The Dark Matter has not been able to develop the Weak Force, as such, failed to obtain the neutron-proton mutation. That is why this black substance does not produce any kind of radiation at all!!

The radioactive disintegration process results in an almost magical transformation. For example, if an atomic nucleus has got 6 protons and 8 neutrons, the Weak Force will detect this imbalance and will be responsible for restoring order by turning a neutron into proton. Thus, the atomic nucleus will have 7 protons and 7 neutrons, becoming a more balanced and stable core.

If we look closely to the values of the masses of neutron and proton, we can see that these values are not exactly equal. The mass of the neutron is slightly higher than the mass of the proton:

$$m_n = 1{,}674\ 928\ 6 \times 10^{-27}\ \text{kg} \qquad m_p = 1{,}672\ 623\ 1 \times 10^{-27}\ \text{kg}$$

This means that there is a tiny amount of mass missing …

With the changing of the neutron there is the birth of a new particle, almost identical: the proton, but with a particularity that makes all the difference. The neutron is neutral but the proton has got a positive charge.

However, the mutation of the neutron does not end with the birth of this new particle, the proton, in this process there is also a new particle that arrives. The final mutation comes with another particle of small mass and negative charge: the electron.

When the neutron is transformed, not only the mass is broken into two parts, but also it divides the charge into two parts.

Nature only has to satisfy a law of equality and equity of the properties of origin, reflecting the Law of Conservation given by the following equation:

neutron mass =proton mass + electron mass

neutron charge = proton charge + electron charge

Moreover, this type of disintegration is a bit more complex, and what experiences have shown is that the mutation of the neutron ends with the creation of another new particle almost undetectable. The complete process of mutation is the following:

Neutron = Proton + Electron + Neutrino

This new and tiny particle, the neutrino, is a small scale version of the neutron, perhaps in an attempt to keep a line of generation … a descendent. Its mass is still unknown, but is so small that it can be considered as having no mass at all, this value is practically zero. And just like his mother, the neutrino has no charge.

There will be so many Primordial Neutrinos in the Universe as protons and electrons. Rather interesting! So many neutrinos ... why?

All of the primitive neutrons in the Universe, which have triplets of quarks in its constitution, were gradually turning into protons, electrons and neutrinos by a process called Beta Disintegration, and only near that period in time happened the consolidation of the first atoms, the simplest element of the Periodic Table: ^1H, Hydrogen isotopes!

And it is because of this choice of Nature, the sacrifice of the neutron, that the most common element of the periodic table is an isotope of Hydrogen, consisting only of a proton and an electron, which is called more frequently by Protio.

Accomplishing this process of mutation gave the possibility to consolidate these new particles and with it a more complex structure of Nature: the Atom. With the conquest of this step in evolution came another natural phenomenon even more magic ... With the arrival of this new variable, the Universe is no longer what it was. From that moment in time the Universe is capable of generating Electromagnetic Radiation, and with it billions of photons are born!

The photons were not always present in the life-time of the Universe. On the contrary of what we usually find in the manuals of Physics, also these particles had to be created...

Very interesting! The photons were not present at the beginning of the Universe...

However, this primordial Universe was still very hot, filled with high energy, which did not allow the consolidation of these atoms for a long time. Since we were dealing with a relatively dense and opaque Universe, so when the electrons produced photons, these photons would interact strongly with other charged particles, immediately colliding with other electrons and protons. Before they can travel freely in a straight line through space, these photons were immediately absorbed. The atoms were constantly ionized and broken and the beams of photons, the flashes of light, were constantly being emitted and absorbed as a result of the higher average densities of particles surrounding.

As such, by this time, the light intensity was still very poor which reflected in a discrete Universe ... brightly light ... as if it was being illuminated by a simple candle!

In order to form stable atoms of Hydrogen, the Universe had to wait until the temperature dropped enough with expansion to allow the stability energy of the atom, what just happened when the Universe has reached 300 000 years old. It was in this moment of time that a truly amazing and wonderful phenomenon happened in our Universe ... the creation of light! The Universe became transparent to electromagnetic radiation in several ways, and the most beautiful of all ... the Visible Light Radiation!

Data acquired from the Deep Field Radiation of the Universe, a kind of radio image of the primordial age of our Universe, confirmed that the cosmic background radiation was released when the Universe was approximately 300 000 years old. This background radiation is a true register in time, and this comes through space from all directions and arrives at the earth with the same intensity.

One of the most intriguing questions that astronomers make about the Universe is: Based on data issued by the radiation, astronomers virtually can 'see' the Universe, and what they see is a primitive Universe too much uniform and homogeneous.

This situation leads us to a problem that no physicist or cosmological astronomer has found a solution to resolve it. It is acclaimed as the Problem of Uniformity or the Problem of Homogeneity.

Chapter VI

THE PROBLEM OF HOMOGENEITY

"If the facts do not fit in the theory,
then you should change the facts."
- Albert Einstein -

At this time, 300 000 years after the Big Bang, the period of inflation had ceased long ago, and this extreme uniformity of the primitive Universe is assumed but not explained.

The obvious next question is knowing how could stars and galaxies arise and evolved from a uniform and homogeneous dense gas of matter, if all particles and matter existing at that time were spread very evenly, distributing in equal ways in space, as if they were forming a velvet tissue!?

From these facts emerges the question of homogeneity. Because its apparently impossible to understand how come a Universe where matter is distributed in a uniform way, could change dramatically and start to form concentrations of matter, stars, clusters of stars, galaxies and clusters of galaxies, and even planets ... which is the current state of our Universe. All of that assuming that the gravitational influence would also be equally well distributed.

Simulations made by computer, insert such data, that is, matter and gravitational influence, and simply conclude that this development would be nearly impossible to achieve. This infers that the formation of stars, planetary systems and galaxies would be, at least, very unlikely, almost impossible and with a very low probability of happening in the natural path of evolution. And this leads us to believe that there seemed to be some kind of 'refining' or a 'cosmic tuning'!

Some might think that the reason why our Universe appears to be this way must necessarily have the intervention of God's hand ...

Let us look more closely at our data and see what is wrong in this simulation.

Track number 2: We only have two variables: primordial matter and gravitational influence.

I leave you to think for a while...

If you were aware and with some attention, until now, I did not have to introduce or talked about any Gravitational Force to explain the Natural History of the Universe!

For some while the matter has remained very evenly distributed, even after the period of inflation, and in this high energy Universe never occurred any kind of density fluctuations in order to form small concentrations of matter because, at the beginning of the Universe, there never was a Gravitational Force!

There was someone who once said:

"If the facts do not fit in the theory, then you should change the facts."

- Albert Einstein -

But then, after some time, matter began to condense and agglomerate, forming stars and galaxies ... but we still do not know how did that happened ... we will get there soon.

Chapter VII

THE PROBLEM OF INFLATION

"The limits of science are like the horizon;
the more we approach them, the more they retreat."
- Bacon -

Initially physicists had supposed that the Big Bang would have occurred as a huge explosion that released an enormous amount of energy. But this explosion did not explain the existence of Uniformity subsequently verified by the cosmic background radiation. To justify this fact Alan Guth had the idea of introducing a new concept, the Inflation. This ultra-rapid inflationary expansion would be responsible to mix evenly all the Universe, producing as a result a very homogeneous distribution of matter. The theory of inflation is needed to explain the initial homogeneity of the Universe, otherwise the Universe would not have had enough time to standardize this uniformity pattern. The era of inflation led to a strange form of the Universe, so that this period has raised many questions and is still waiting for some clarification. There are some concerns that a new period of inflation might happen again.

Cosmologists seek to understand how come during the first fraction of a second of the birth of our Universe, immediately after the explosion of the Big Bang occur, the expansion of the Universe has achieved extremely high values, exceeding even the speed of light!

During this period, the observable Universe, the very fabric of space and time, has greatly extended in much higher values than those calculated today on the expansion of the Universe. During the period of time that lasted for inflation, the Universe doubled in size every 10^{-35} seconds. Doubled up, so hundreds of times, expanding its volume at least 10^{50} times, while the temperature fell vertically, from 10^{28} K to 10^{23} Kelvin.

Will we be able to explain the reason why this period of inflation had to happen? And also why it has ceased?

Specific data show us that the period of inflation did not occur simultaneously with the explosion of the Big Bang, but a few fractions of seconds later.

As you can see by this chart that shows the evolution of the radius of the Universe immediately after the Big Bang occurred, it appears to be that the initial explosion was, at the beginning, very regular and linear. It is after then, in a specific moment in time, that the expansion becomes inflationary, as we can see by the following graphic.

This graphic of the inflation era show us the relation between the radius and the temperature of the Universe ... I wonder what has unchained this inflation?!...

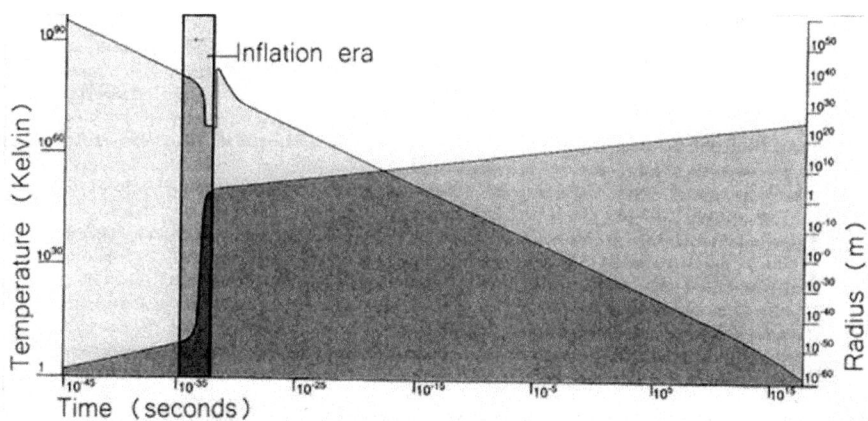

- Inflation Era -

This period might seem strange to understand. But, perhaps, it may not be so strange as we think ...

Once again, based on the assumption that at the beginning of the Universe there were no Gravity Forces, that is, assuming that initially there would be no gravitational influence, thus the initial explosion would have started to expand in a regular way, but without any element of contraction exerted by any gravitational forces, this explosion would certainly expand continuously.

Let us now make an analogy: When a gas is expanded, the relative density of the gas ceases to be constant, because the particles of this gas are redistribute as the volume increases. And the density, which means the amount of particles present per unit volume, is being increasingly smaller as the gas becomes more rarefied, more empty, until is filled practically by vacuum.

In the early expansion of the Universe there was a similar process. Removing this variable, the gravitational influence, the Universe was free to expand forever, at its free-will and very quickly. While the density was reduced dramatically, reaching a point where the density per unit volume was almost zero. If this expansion does not ceased, the barrier of zero is exceeded and the Universe reaches a negative density. This negative density leads to a very critical moment: This transition shows the power of the False Vacuum.

Once the Universe enters and reaches this false vacuum, opens the door to an enormous amount of energy. It is this huge amount of energy that enters and unchains the inflation.

But our Universe does not allow negative densities and therefore the repulsive energy of the false vacuum is unstable and decays very quickly. The door that was opened must be quickly closed. Once there has enter enough energy to restore a positive density, for values just above zero, the inflation ends, stopping automatically.

The period of inflation lasted very little time, a few micro seconds ... fascinating!

Track number 3: The absence of Gravity leads to negative densities.

Cosmologists can be relaxed, there will probably be no more period of inflation in this Universe.

21 SOLUTIONS FOR 21 QUESTIONS
OF THE XXI CENTURY PHYSICS

PENÉLOPE
FOURNIER

Chapter VIII

THE PROBLEM OF CRITICAL DENSITY

"Imagine the Universe, beautiful, just and perfect.
Then, ensure you of one thing: The 'One',
imagined it a little better than you did."
- Richard Bach -

Another issue currently open is a paradox raised by cosmology, also much too difficult to understand and very peculiar. It is the problem of the average density of matter in the Universe, or more simply, the problem of Critical Density.

The fate of the Universe is dependent on its density. To understand this problem we must consider that the destiny of the Universe is dependent of a precious number omega: Ω. This number represents the overall average density of all the matter / energy in the Universe.

If the density is high, the influence generated by the gravitational forces are stronger and they will stop the expansion, forcing the Universe to close and collapse, getting back to a Big Crunch; otherwise if the average density of matter is low, nothing will be able to stop the process of expansion, and our Universe will expand forever, becoming in to an open Universe, dispersed, empty and cold.

What occurs in practice is that the density of our Universe has got a very delicate value, very close to the critical density, which means $\Omega_0 = 1$. This number puts us exactly between an Open Universe and a Closed Universe, which makes our Universe with a radius of curvature practically plane, flat and null.

These data report that the density of matter/energy existing in the Universe is not substantially higher or lower than the critical density and therefore the space is not substantially curved, positively or negatively.

A real puzzle that points to the fact that our Universe is neither open nor closed, but somewhere between the middle of these two states.

Within an infinite number of possibilities that could have led to an open or closed Universe, how come our Universe seems to have exactly the critical density Ω_0?!

This issue is very delicate and it's often associated to another 'cosmic tuning'!

Currently, it is assumed this incredibly precise number of matter/energy and density of the Universe as a fact, however, still waiting for an explanation.

The density and the fate of the Universe are closely related according to this variable omega:

$\Omega_0 = 1 \Rightarrow$ Plane Universe

$\Omega > 1 \Rightarrow$ Closed Universe

$\Omega < 1 \Rightarrow$ Open Universe

The observed abundance of Hydrogen, and also Helium indicates that the baryonic density of normal matter cannot be greater than 0,1 of critical density. However, to this value we must add the dark matter, which in total will not exceed an overall density much higher than 0,2.

This implies that the density of our Universe is very close to the critical density. Why is omega so close to 1? Is it really equal to 1?

Given that the Universe has been expanding and in a continuous process of creation since the period of inflation, we could consider the fact that the current density is very near to the critical density implies that this constant would have the exact value equal to 1 at a time closer to the beginning of Universe.

This leads us to suspect that the initial critical density of the Universe would be exactly equal to the critical density. Which means that at sometime in the past $\Omega_0 = 1$!

With the new model of inflation presented, with absence of gravitational forces, this issue could be easily resolved and this cosmological problem of the critical density would no longer be a problem!

After this period of huge inflation expansion occurs, whatever it was the geometry of the Universe before this period, the new geometry would be necessarily flat after the expansion; and the value of its density would be necessarily 1.

Track number 4: The Universe could arise with any density it desires ... The period of inflation ends precisely when the Universe reaches the critical density! As soon as the energy of the false vacuum enters and restores the positive density, just above the value zero!

There is a specific process and a scientific principle that explains why the density of the Universe has the value it has. This number of this mysterious constant did not need any 'cosmic tuning'!

Thus, by changing only one variable, we can solve three major cosmological problems! In the hypothesis used it was sufficient to remove from our equation one single variable: the Gravitational Force!

Chapter IX

WHAT IS THE FALSE VACUUM

"If our minds were free of training and conceptions,
the illusion would not occur and the true mind would be free
to see everything."
- Tao-Shin -

The mystery of the false vacuum is also fascinating!

The quantum fluctuations of the false vacuum show us a huge amount of energy, opening a door into the unknown...

Thinking in abstract, we all know that oil and water do not mix, because they have different kind of densities.

Imagine that we were all living in a bubble of oil. We could be completely surrounded by water, and this substance could be so very close from us, but we would never know anything about it!

Our Universe is just a bubble of oil merged into a huge ocean of this hyperspace...

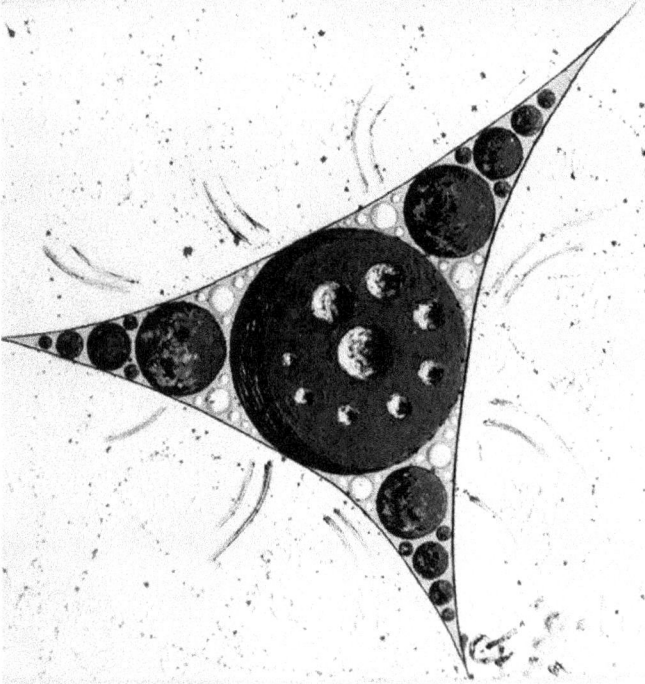

- Oil painting: " Many Worlds " C.P.F. -

Chapter X

WHERE DOES GRAVITY COME FROM?

"Life is like a sculpture. It is a question of being capable to see what others do not see and then, later, with a chisel, take of what surplus."
- Michelangelo -

With this new model set for a Universe with no primordial gravitational forces, many issues would be resolved.

In order to achieve the resolution of these issues we must abandon the preconcept that considers Gravity as an innate force of the Cosmos and inherent to masses!

By introducing this revolutionary change in the field of Physics, another aspect should be focus, which is:

What I think is that the intimate structure of matter is not in any way related to the Gravitational Force! And consecutively, the gravitational force is not directly related to this physical variable: quality or quantity of Mass!

It is clear that this new perspective would, necessarily, raise another question. After all, where does Gravity come from?

This is the big question that reunites us here today!

As we can see in several books of Physics, when they present several important dates of the Natural History of the Universe, we can see that the Force of Gravity appears at the beginning of time, together with the birth of the Universe, at the moment of the Big Bang. The reason for that is, apparently, somehow physicists consider that Gravity is intimately related to time.

Referring a small excerpt, it appears the following:

63

'In the first moments, when the Universe was only 10^{-43}s of age, soon after the explosion of the Big Bang, space and time were still being created. The forces of Nature were combined in a single Primordial Force, designating it by the Big Unified Force. That specific period is designated as the Planks time, and its details cannot be explained because we need a theory of Quantum Gravity (...) and at this time all the forces were still being formed. "

According to what has been demonstrated, until now, we did not need to include any Gravitational Force to explain the Natural Evolution of the Universe.

I almost dare to believe that maybe we don't need any Quantum Theory of Gravity!

Clearly, this deduction would, automatically, turned many physicists bold and in a state of shock! What would not be very convenient or appropriate.

Before getting into fast conclusions, let us start, once again, from the beginning...

It would be a good start recalling that the concept of Mass and the concept of Weight of an object are two very different qualities in Physics. It is important not to confuse them.

We know that our weight on Earth does not have the same value as the weight in the Moon. Our weight on the Moon is less than the weight we feel on Earth. And when this happens it is not because it had occurred a deficit of mass. The Mass remains the same, because Mass is a measure of the quantity of matter, number of atoms, but the weight is different, because the weight is a measure of a force applied to an amount of mass.

Our weight in the Moon is smaller, simply because mass feels less weight, because there is less gravitational forces acting.

And with this I wish to conclude the following: that the mass can exist even in the absence of gravitational forces, simply, it ceases to feel weight ... very interesting!

The weight depends of the place where we remain, of where we are, which means that it depends of the local magnitude of the acceleration caused by Gravitational Forces.

Correct?

If you agreed, then, you will also agree that Mass and Gravitational Influence may be independent concepts, and then we can conclude that they are not directly related!

The gravitational influence only transmits weight to the masses. This effect shows up only when combining these two variables: the amount of mass exposed to a gravitational field, and as a result we will have heavy matter.

Although the mass of an atom is concentrated in its core, in the nucleus, we are deluded by Nature and we believe that the Gravitational Force also comes from this center. We do immediately implement the following deduction: obviously if the higher concentration of gravity is where it is the largest concentration of matter, nothing more than logical to conclude that Gravity comes from this quality of Mass.

But as we have seen before this quality of Mass it is not directly connected with Gravity!

But still we try to explore, at all costs, what is the secret of matter, what is its minimum indivisible composition, and so we hope to obtain the Quantization of Matter and achieve a Quantum Theory of Gravity. This is the biggest enigma that draws the attention of thousands of physicists around the world...

I'm sorry to disappoint you but, although the largest concentration of mass leads to higher values of gravity, which is inside the nucleus of the atom, the source of the gravitational field, this one, it does not come from the center!

To reach into this conclusion it would be very simple, we just have to go backwards in time and continue with our Natural Evolution of the Universe.

Where were we? ... Ah! ... Yes ... 300 000 a. B.B. (after Big Bang)

Continuing with our model of absence of gravity in this original Cosmos, we went back in time and enter in the period of Uniformity.

This Uniformity who appeared and remained even after 300 000 years after the period of inflation occur, could be easily explained if we assume that nothing could perturb the position of a single particle, if we consider that there are no gravitational forces acting on mass and so nothing could undo this uniformity and cause the minimum change of density or concentration of matter.

Thus, the second law of Newton's Dynamics, the Law of Motion, or F = m.a, could not come into action, because the particles would not be attracted by anything, because if no force is being applied to them, then the primitive matter would remain exactly in the same initial positions.

Except for the relative distance between them, that would increase due to natural phenomenon of the Universe expansion and extension of space itself. An interesting phenomenon, the distension of space observed by Hubble, when he confirmed that all the galaxies were being away from us at the same time ... very interesting ... but the explanation of this phenomenon we will leave for some other time.

So far, as we saw, what happened in this Cosmos has been the development of matter itself. The transformation of primordial particles into primitive nucleus of quarks and that some of these primitive nuclei obtain the neutron-proton transformation, but not all of them. Most of them remained as Dark Matter and only a small part evolved into Chemical Elements.

All the Dark Matter that astronomers have detected in space only represents all the matter that has failed in the natural process of evolution of the Universe and so therefore is very unlikely, and extremely difficult, that the first atomic nuclei were formed when the Universe was only three minutes old!

With this new scenario we must note that although all the amount of matter is evenly well distributed, not all of it evolved in the same way. Only a small part, some of this isolated nucleus, randomly distributed, were able to form Hydrogen isotopes.

From this moment in time when a familiarly atom of the periodic table is formed, consisting of one proton and one electron, it also emerges the Electromagnetic Radiation, which begins to fill the entire space.

The evidence of this cosmological event in time is shown in the emergence of photons detected in the Cosmic Background Radiation.

Before this period in time it is impossible to obtain any 'visual' fossil registration of our Universe.

What I intend to say more directly is that, it was approximately during this period that the Electromagnetic Radiation is formed! Also this Force had to be created and born from a natural process of evolution! Without protons and electrons there is no Electromagnetic Force.

Again, it should be noted that the Electromagnetic Force arises only as a result of the interaction of photons and electrical charges.

This new model of our Cosmos aims to focus a natural Evolution, very wised and spared. Nature only needed to create one Force at a time ... one after another ... or almost!

And with this, I have said almost everything!

If you noticed, what happens after this moment in time is that the Universe begins to evolve in a very different way. The first concentrations of matter began to emerge, more specifically, concentrations of Hydrogen gas, which can only be explained if we assume that there are gravitational influences. Thus, concentrations of matter can only occur if we admit the existence of Gravity. Somewhere during this time emerged the Gravitational Force!

We know that Gravity is the weakest of all forces. Its action is slow but very patient.

The first evidence of large concentrations of matter occurred thousands of years after the Big Bang in the form of Quasars very dense and hyper-luminous.

These early stars are extremely interesting. With the size of a single star they can produce powerful radiation, the same as an entire galaxy. These Galactic nucleuses produce an intense quantity of light, so high that the density of photons emitted at this time per unit volume was an order superior then the present.

The quasars are objects distant in time and as such they no longer exist. These old stars are the fossil registrations of the first steps in the evolution of our Universe involving high energy, and these primitive concentrations of mass are the first evidence of the formation of large structures of matter in order to form the first stars and the first galaxies. The time required to achieve this amount of matter is actually huge, probably of several million years.

Interestingly, if we imagine this process backwards in time and reverse the process of gravitational attraction that led to the concentration of such matter into Quasars we probably arrived exactly about the same time of Uniformity.

In deed the time required to cluster this quantity of matter is really huge.

Obviously, this leads us to conclude that the Gravitational Force must have emerged very early.

We could assume that the gravitational force emerged sometime during the period of Uniformity or Homogeneity... again!

Thus, we could also assume that within this period, approximately 300 000 a.B.B., as soon as a single Atom of the periodic table was build, suddenly emerged the Electromagnetic Force produced by electric charges, and also the Gravitational Force produced by ... ... well ... let's say for more complex compositions of matter.

With this we can conclude that only the common atoms of the Periodic Table are capable of producing Electromagnetic Radiation and Gravitational Radiation ... It is on purposes, I consider it as Gravitational Radiation.

This was the true Era of Radiation!

To complete the pack of our forces, there is one still missing: the Strong Force. What is the Strong Force? We know that the Strong Force is responsible for keeping the protons in the nucleus together. Without the presence of this force equal charges would repel. But why would Nature need to create the Strong Force?

The answer to that is very simple. It is confirmed that Nature has a tendency to form groups of matter and clusters of mass with larger numbers of atoms becoming a structure more refined.

In order to acquire a higher degree of complexity, the union developed in order to form atoms more complex than Hydrogen. The next effort followed in an attempt to build atoms of Helium.

Form an atom of Helium was not easy. It was necessary to maintain two protons very close to each other, concentrated in a nucleus and two electrons orbiting around, near the core, in the periphery of the atom.

Using again the registrations of our Cosmological Almanac, it appears that, almost since the beginning of the Universe, there are 15 000 million years ago, which has been happening the formation of Hydrogen, so that this atom is today the dominant chemical element and the one which exists in higher quantity.

Soon after this, we have the Helium, whose relationship of proportion between them is 75% Hydrogen and 25% Helium. However, from this 25% amount of Helium only 10% were formed in the stars, the remaining 90% are prior to the formation of stars. What studies have showed is that this chemical element has evolved relatively soon, long before the birth of stars.

Of all the constituent elements of the Periodic Table, in our Universe, 99.9% are Hydrogen and Helium and the remaining 0.1% represents all the other heavier elements like Oxygen and Carbon, etc.

The proportion of Helium is still very high ... curious...

The Helium is a strong union of two atoms of Hydrogen...

Do you read my thoughts?

First, the Electromagnetic Force came with the Hydrogen atoms, with the creation of electric charges;

Second, the Gravitational Force also arises with the formation of Hydrogen atoms, with the creation of more complex structures of matter;

Third, the Strong Force is already present soon in the life time of the Universe, an example of its manifestation is in the formation of atoms of Helium, but their percentage is lower than Hydrogen because, first, it was waiting for the gravitational force to pull and concentrate some amount of Hydrogen;

Fourth, the Strong Force has possibly arisen simultaneously with the Electromagnetic Force and the Gravitational Force!

Very interesting...

First conclusion: everything returns to the same moment in time: the formation of the Atom!

At first glance this may seem an obvious conclusion, or most obvious, but perhaps it is not so clear as it seems.

With the formation of the atom all the other three forces of Nature were also formed. Therefore, they did not always exist. Also they had to be created in some moment in the evolution of the Universe!

This already goes against the principles of many physicists, who believe in a more Biblical Physics than a Natural one!

The theoretical physicists assume that all the four forces of Nature were already present at the beginning of the Universe!

As same as the Bible that assumes that humans were already present since the beginning of the formation of the earth, that believes man evolved from Adam. Thus the physicists also believe that all fundamental forces of Nature were all born at the same time, emerging all from the Big Bang!

In my view, this is a critical error...

Although all physicists agrees that radiation and photons are innate properties of the Cosmos, as well as all the Forces of Nature emerged with the Big Bang, the truth is that the four forces of Nature did not emerge all at once and at the same time in the form of a Grand Unified Theory!

Continuing with our Natural History of the Universe, we could assume that the three forces of Nature - The Electromagnetic Force;

Gravitational Force and the Strong Force - were created at the same time, simultaneously with the formation of the Atom.

We could speculate that they might have had a common cause.

We could speculate a little more and believe that they could perhaps share the same origin!

But how would that be possible? Is there any similarity or any chance of a link between these three forces, supposedly so different? Clearly, these three forces of Nature have distinct functions and interactions! But is there any chance of connection between them?!

If we could discover what relates these three forces, which is its common gene... we would have the key to our Pandora's Box, and with it, we would find out everything!

Chapter XI

THE ORIGIN OF THE FORCES OF NATURE

"We are just dust from the stars."
- Carl Sagan -

According to the Standard Model of Cosmology, scientists in general, accept that there was a moment of creation for the Universe. In this model, physicists proposed a Unified Theory in which they believe that in fractions of a second after the Big Bang occur all of the four known forces of Nature already existed, and that these individual forces were, at this period in time, combined together in the form of a single large super powerful Force, known as the Great Unified Force.

The origin and characteristics of this unique Force is, however, very uncertain and undefined, but it is considerer that this was the master Force responsible for the first moments of our Primitive Universe, which in this initial stage of evolution was not formed and filled by matter but only by energy in the form of radiation.

The theory assumes that all of this four forces, with very distinct interaction and functions in Nature, the Gravity Force, the Electromagnetic Force, the Strong Nuclear Force, and the Weak Nuclear Force, were already present since the beginning of the Universe and that all of these four forces acted in a unique way in a form of a Great Force, with mysterious and natural properties, but, unfortunately, not much is known about the function and behavior of this force...

Force of what?!

Scientist anticipate that this Unified Force change its characteristics and properties over time, and as the Universe began to cool down this

Great Force was separated and broken, gradually branching into the four forces currently known. However, it is not given any evidence sufficiently clear that justifies and explains this process. Physicists in general just do believe that this was the way that happened. Their arguments are based on the coupling constants of these forces and in the union and convergence of the same when framed in a High Energy Physics, characteristic of a Primordial Universe.

Without having to appeal to very refined calculations, perhaps there may be another solution. The analysis that it will be presented is based on a much easier version and follows from the cosmological facts and evidences.

Returning to our Natural History of the Universe, it would be good to make a pause for a moment and reflect on the following question:

What is the common characteristic of all the forces of Nature?

Looking like this, suddenly it appears that we cannot see any similarity between them, in anything. These four forces seem to have nothing in common ... or almost nothing!

In order to try to make a clear identification of these forces, we will try to recapitulate the properties and characteristics of each one of them in detail. Let us began buy investigating if there is any relationship between the Electromagnetic Force and Gravitational Force.

The first thing that we learn in school is that the Electromagnetic Force is a force of radiation and that the Gravitational Force is not a radiation force.

Please do get ready for a new journey in time, because not everything that we learn in school is correct!

Let us now make some comments about these two forces.

At first, we should note, the enunciation of the Universal Law of Gravitation written by Newton: "Two bodies with mass feel a force of attraction in the direct ratio of their masses and in inverse ratio of the square of the distance between them."

Take now the statement of Coulomb's Law for the Electromagnetic Force, "Two electrically charged bodies exert a force proportional to its charge and inversely proportional to the distance between them." In this case, the law has something more to add: if the charges are opposite, there will be attraction between them, otherwise there will be repulsion.

First, we should note the similarity in the structure of these two laws: both say that a force is proportional to a specific attribute: mass in the case of gravity, electric charge for electricity; both agree with a elementary constant of the environment, 'K' dielectric constant, and 'G' gravitational constant; and both vary in the inverse ratio of the square of the distance. The two formulas of these forces have in fact a similar pattern:

$$F_{em} = K. Q.Q / r^2$$

$$F_g = G. m.m / r^2$$

Is this an accident of Nature? Nature has got very few accidents!

It was based on this symmetry of Coulomb's Law and Newton's Law that Einstein think that this could not be a pure coincidence. Until the end of his life, Einstein tried to describe all the forces of Nature through a unique formalism of unification. He took years to try to carry out this unification which could describe all the forces of Nature through a single equation. He always believed, until his last breath, there was a relationship between these forces.

Unfortunately, he did not have the possibility to confirm that there is a relation! Einstein was right ... once again.

The truth is all written in the Great Book of Nature!

Let us highlight the assertive characteristics of these two forces:

1st - In the case of Electromagnetism identical poles repel each other.
   (Electromagnetism is a naturally repulsive);
   In the case of Gravity, identical poles attract.
   (Gravity is a naturally attractive).

2nd - In the case of Electromagnetism there is always an electromagnetic dipole.
   (There are no magnetic monopoles, when a magnet is broken we will always get a new magnet with two distinct poles);
   In the case of Gravity, the opposite occurs, there is always a gravitational monopole.
   (There are no gravitational dipoles; when a mass gets divided we do not find 'negative' masses that experiment gravitational repulsion).

3rd - What one Force has got, the other does not have;
   What one Force does, the other one does not do.

Is it possible to achieve a pattern from this information?

I see a pattern. You do not see the pattern?

Here there is no Symmetry, not at all ... that is a fact ... but there is something else ... what we have here is Asymmetry!

It seems that there is hidden in these laws some kind of an Algorithm for Asymmetry!

Is it possible that Nature also understands a little bit of Informatics?!

Symmetry, symmetry ... when I pronounce this word, does it not make you remember anything?

Asymmetry … does it not remind you of something?

... Asymmetry... Asymmetry … Asymmetry!

It was precisely at this moment that my brain was flashlight, and all my neurons were turned on in flash!

Magnificent, superb, simply phenomenal!!

Another clue: A new tool that physicists invented to study the physical properties of Nature is called Symmetry.

As you all know, the ordinary concept of symmetry is related with the reflection of an object in front of a plane mirror. An object is related to its image in the mirror. In a symmetrical object, a sphere for example, the reflection shows the same characteristics as the original object and even if we try to make any rotation in the original object, its image in the mirror does not change. This means that this transformation has not led to any differences in the image analyzed. The final image appears the same. And this is considered as an example of Geometric Symmetry, however, there are other types of symmetry, more abstract, used by many particle physicists.

The concept of symmetry applies to certain proceedings in the field of Physics. It´s technical term is called of Parity or simply Symmetry. The symmetry is important in many physical processes of Nature, the break of the symmetry too.

Apparently all the forces of Nature respect the symmetry, but not one of them … that breaks it... What is the force that is associated with asymmetry? What is the force that does not respect the symmetry in the mirror? I leave you to think...

Let me reformulate the question. With this scenario that we have seen so far, I ask you:

What could be the Mother Force of all forces? The Force that was always present, almost since the beginning?

Well, of course! Certainly we are all referring to the Weak Force!

And what is the Weak Force?

It is the force responsible for radioactive disintegration. Correct?

Well, then we can also say that this is the Force of Radiation!

And don't you think that a Force of Radiation would produce new Forces of Radiation ... a little bit more weak ... but still forces of radiation?

I think that what we need to consider is how, at this scenario, could that be possible. Let us try to reveal the process of this mechanism.

Take the example of the neutron-proton mutation or, more precisely, the Beta disintegration.

Nature can not invent. But for example, if I have one orange, I cannot invent two oranges, but I can divide my orange. Right?

In the process of Beta disintegration Nature cannot invent, therefore it divides their properties. In the specific case of neutron-proton mutation what we have is the following:

1st - Division of Mass:

neutron mass = proton mass + electron mass + neutrino mass

2nd - Division of Electric Charge:

neutron charge = proton charge + electron charge + neutrino charge

We know that there are particles mediating the Weak Force which are responsible by causing this conversion. These mediators are Bosons W^+, W^- and Z^0.

Exactly three mediator particles … Very interesting! In its physical notation we have:

$$n^0 = W^+ + W^- + Z^0$$

$$n^0 = p^+ + e^- + v^0$$

Nature does not reveal immediately all of its secrets! Can we get from here a third division?

3rd - Division of …

What if I change the equation in to this way:

$$\text{Weak Force} = m_p^+ + m_e^- + m_v^0$$

And if I remember I need to know where does this come from:

Strong Force + Electromagnetic Force + Gravitational Force

And if I rewrite the equation in to this way:

Weak Force = Strong Force + Electromagnetic Force + Gravitational Force

Already seeing the pattern? The Weak Force divides the Mass … the Weak Force divides the Electric Charges and…

The Weak Force divides the FORCES!

The Weak Force is weak only by name, because it is the central structure of all the Forces of Physics!

Chapter XII

GRAVITONS LOCATED?!

"Nature does not do anything in vain."
- Aristotle -

If we relate all the forces of Nature in a common equation, it arises the following interpretation:

$$F_{Weak} = F_{Strong} + F_{Electromagnectic} + F_{Gravity}$$

$$F_w^0 = F_s^+ + F_{em}^- + F_g^0$$

$$n^0 = p^+ \quad e^- \quad v^0$$

From this relation we can conclude that the mediators of the Forces of Nature are:

bosons = gluons + photons + neutrinos

So many Gravitons on the loose! Disguised as Neutrinos ... the perfect dissimulation!

Only like this everything fits very well! And everything stays so organized, so simple and beautiful!

The Theory of Gravity requires the existence of these Gravitons with a very specific and a very defined interaction with matter. As such, these particles should be everywhere has a reflection of the interaction between masses. Therefore, they were represented by a mysterious particle, not yet discover and never detected: the missing graviton!

This graviton should be a stable particle; electrically neutral; probably with very little or no mass at all; and be present in large number of quantity, distributed almost uniformly throughout the Universe ... which are exactly the characteristics of neutrinos!

The fact that there are so many neutrinos in the Universe cannot be a coincidence of Nature! What is the role of so many neutrinos? If we look close, until this moment we did not give any fundamental role to these specific particles so subtle and omnipresent.

So many neutrinos … just walking around! I believe that Nature does not do anything in vain.

The number of these particles is really immense, and is about equal or even greater than the number of photons in the Universe, and both travel at the speed of light, which is the main feature of the mediators of the interactions.

The neutrinos are, very likely and most simply, the most abundant particle in the Universe!

How come, such a fundamental particle can pass in front of us and stay so unnoticed?!

We are constantly being bombarded by billions of photons and neutrinos, or gravitons ... as you wish. These are the particles responsible for transmitting to our atoms the Electromagnetic interaction and the Gravitational interaction.

The Nature exceeds by simplifying its laws!

Chapter XIII

GRAVITATIONAL RADIATION

*"If you have an idea and if at first glance it does not
look like completely absurd, then,
there is no salvation to it."*
- Albert Einstein -

What kind of force is Gravity?

If we aspire so much to obtain a Theory of Everything, the Great Unification Theory, the Cosmic Unity, we must overcome our prejudice that establish Gravity and Electromagnetism as two distinct and quite completely different forces. In my point of view, they are not so divergent.

Considering these two hypotheses:

That Gravity is not directly related to mass; and that Gravity is a force of radiation.

Is it possible that we can formulate a new Theory of Gravity based on these assumptions?

1st Scenario:

If it does not come from the center of the nucleus, from the concentration of mass, then, where does Gravity come from?

The Gravity emanates from everything: from matter, from heat, light, even from the very Gravity itself!

The light, feels weight, therefore it is affected by Gravity. Let us remember that a photon it is a particle which has got no mass! A photon is not only diverted of its trajectory by the presence of a Gravity field, due to the proximity of huge masses like stars for example, but it is also able

83

to attract other objects. A ray of light with sufficiently high energy would attract us, or vice versa ... it is relative.

How come a photon, a particle with zero quantity of mass, can be affected by Gravity and can also produce a gravitational field!?

The movement can also feel weight. It is known that a star with a faster movement of rotation exerts a gravitational pull stronger than an equal star with the same mass but moving in slow motion ... the reason why this happens it is not because it was added more mass ...

The Theory of Gravity describes that mass can produce a gravitational field and that from this, depending on movement, more Gravity can be produced.

Gravity can produce Gravity and so on!

Gravity can also be generated by a magnetic field in movement around a Space Station! ... Very interesting... very interesting indeed!

Does Gravity also suffer from a syndrome of personality disorder? What is the only variable in these three cases? For sure it is not the Mass the correct variable...

After all, where does it come from so much quantity of Gravity?!

We are constantly saying that the Force of Gravity has got a completely different structure from other forces;

I often listen that Gravity is a Classical Force, completely distinct from all the others;

Many times is said that, indeed, there is no Force of Gravity at all, in fact this is only and simply just a geometric property of space-time itself;

I constantly hear if Electromagnetism and Gravitation have something in common or any kind of similarity, it is only because their strength varies with the inverse square of the distance!

This is very annoying! I feel the obligation to intervene.

I consider that all these characteristics, connotations, and descriptions are quite disconcerting.

Point n. 1 - Both are field forces;

Point n. 2 - Both vary in inverse ratio of the square of the distance;

Point n. 3 - Both have infinite range;

Point n. 4 - Both propagate at the same speed.

Curiously, the speed of propagation of Gravity it is not instantaneous, its speed is equal to 'c', the speed of light! Is this a coincidence?

The Universe has many little coincidences!

Can it be that Gravity it is also an Electromagnetic Wave?!!

Some might think this is a very strange concept, but if we want to address a new Physics, we must demand new ideas.

Only because it is suggested and of common agreement that Gravity is not a force of radiation, is now for me much more difficult to say the opposite and present Gravity as an electromagnetic wave.

Let us start from the beginning:

The biggest evidence that Relativity Theory has reported was the equivalence between gravitational mass and inertial mass, thus, what this means is that Gravity has an origin similar to inertia and, as we know, the inertia is closely related to movement.

Let me try to relate the following, speaking without much accuracy, we can say that:

At rest, electrical charges can only produce an electric field, this is:

ELECTRO-STATIC FIELD

In motion, electrical charges can produce electric and magnetic fields, so we have:

ELECTRO-MAGNETIC FIELD

The dislocation of the electric and magnetic field combined together produces a new field, the gravitational field, which can be designated as:

ELECTRO-MAGNETIC-GRAVITATIONAL FIELD

Summarizing, in this first approach, we can just stay with this general idea which says that:

With the dislocation of a field we can always define a new field. The movement of the electrostatic field produces a new field, the electromagnetic field. The dislocation of the electromagnetic field also produces a new field, the gravitational field!

The dislocation of all of these fields moving in syntony is consolidated with the formation of an Electro-Magnetic-Gravitational Field, which requires the production of Gravitational Waves!

The unification of the Electricity with Magnetism has caused a great discovery: the formation of Electromagnetic Waves;

The unification of Electromagnetism with Gravity also has a remarkable implication: the formation of Gravitational Waves!

At this moment, I am reminding me of a quote from Einstein: "If you have an idea and at first glance, it does not look like completely absurd, then there is no salvation to it." - Albert Einstein -.

Take then a more practical example. Let us remember the diagram of the Electromagnetic Spectrum and its natural radiation.

From one extreme to another there are several known radiation distributed according to their frequency.

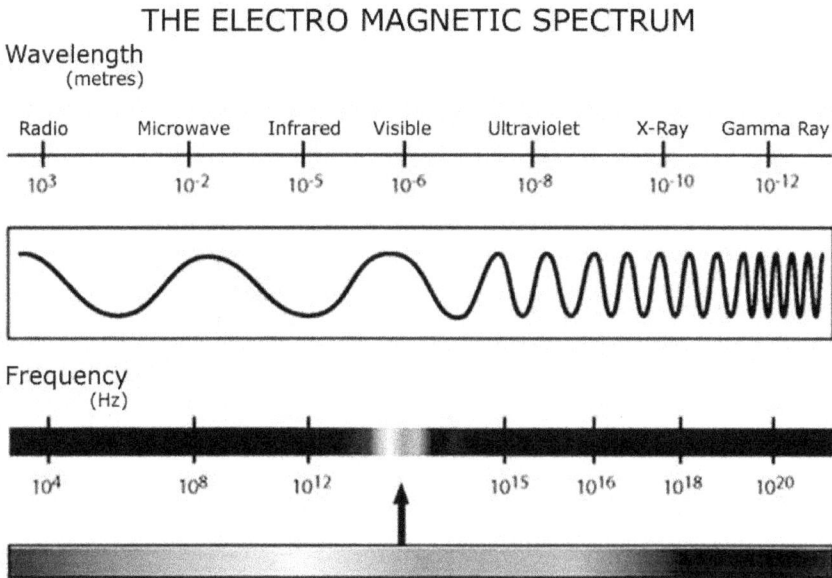

THE ELECTRO MAGNETIC SPECTRUM

Wavelength
(metres)

Radio	Microwave	Infrared	Visible	Ultraviolet	X-Ray	Gamma Ray
10^3	10^{-2}	10^{-5}	10^{-6}	10^{-8}	10^{-10}	10^{-12}

Frequency
(Hz)

| 10^4 | 10^8 | 10^{12} | | 10^{15} | 10^{16} | 10^{18} | 10^{20} |

- Electromagnetic Spectrum -

As you can see in this spectrum it does not appear any gravitational waves. This is where the adversary makes the mistake!

As you can see, at the beginning of the spectrum (right side) we have the highest frequency of all: the Gama radiation with a frequency close to 10^{20} Hz; after that we have the X-rays 10^{18} Hz; Ultraviolet 10^{15} Hz, at 10^{14} Hz the Visible Light radiation, and then we move towards the lowest frequency: Infrared 10^{12} Hz, at 10^{10} Hz we have the Microwaves and Radio Waves which can be extended up to 10^4 Hz or more and ... does the spectrum ends here?!

As we know, a package of energy corresponds to a quantum of energy. The value of this quantum depends on the frequency of light, which is given by the equation E = h.f   The greater the energy carried, the greater the frequency of radiation. The higher the frequency, the lower the wavelength. Similarly, the lower the frequency, the greater the length of the wave.

Lets us make a simple attempt to estimate the frequency of a Gravitational Wave, through the equation E = h.f

But first we need to know what is the energy of a Gravitational Field.

In order to make an approximate estimation for this value we can make use of the relative intensity of the forces that we know. The strongest force of Nature is the Strong Force with the greater intensity, next on this scale we have the Electromagnetic Force, 137 times weaker - a very interesting number - then we have the Weak Force 10^6 times weaker than the Strong Force, and finally the weakest force of all and at the final of the spectrum comes Gravity 10^{40} times weaker than the Strong Force.

With this information, we can say that the magnitude of Gravitational Energy has got an absolute value about 10^{-40} or $1 / 10^{40}$.
Doing a simple calculation...

$$E = h.f \iff E/h = f$$

$$\iff f = E / h$$

$$\iff f = 10^{-40} / (6,6 \times 10^{-34})$$

$$\iff f = 6,6 \times 10^{-6} \ Hz$$

This will be a good approximation to the frequency of a Gravitational Wave, which shows that it is a wave of low frequency, very low. Its frequency is lower than the Radio Waves, and therefore its wave-length must be greater than the Radio Waves.

Let us see if we can get some approximation to the wavelength value.

Knowing that:

$$f . \lambda = 2\pi . c$$

$$\Leftrightarrow \lambda = 2\pi . c / f$$

$$\Leftrightarrow \lambda = 2\pi . 3 \times 10^8 / 6,6 \times 10^{-6}$$

$$\Leftrightarrow \lambda = 1,6\pi \times 10^{14} \text{ m}$$

This means that the length of a Gravitational Wave is very big, very big indeed! We should seek it at the end of the spectrum, just below the Radio Waves ... we will need a very big antenna!

Chapter XIV

ELECTRODYNAMICS STABILITY OF THE ATOM

" Theoretically possible, infinitely difficult."
- John Gribbin -

Most of the contemporary physicists do believe that there are, currently, two types of Physics:

In one side we have the Classical Physics and at the other side we have the Quantum Physics. And what does this mean?

For example, the quantum physicists say that the electrons in an atom have quantum energy levels very well defined and that this explains why the atoms are stable. Then, they consolidate their argument saying that the Classical Physics obeys to none quantum principles. The argumentation and the reason why this happens is because they consider that the electrons are quantum objects and, therefore, they do not share the same properties of Classical Physics.

Please do pay a little bit more of attention in this last sentence. Look well, to avoid the paradox, physicists defined the following:

Accelerated electrical charges in a none quantum level do emit continuously radiation;

But, accelerated electrical charges in a quantum level do not emit any kind of radiation.

Therefore, the physicists do divide Physics in two parts: the Quantum Physics and Classical Physics, with very distinct properties. In my point of view, this is another crucial mistake ... the Physics of this Universe, as far as I know, is unique and only one!

It seems that all objects in the Universe have to be in a circular movement. With a specific trajectory which seems to be responsible for the delicate balance of Nature.

Is this kind of movement, somehow privilege? Some sort of a Golden Movement?!

This circular motion is indeed very interesting. If we look closer, this type of movement seems to be present in all the main structures of the Universe and of Nature. Since the atom, the rotation of stars, the translation of planetary systems, the trajectory of comets and even in the movement of the Galactic disc ... everything has to be in rotation and in perfect harmony with this circular acceleration!

Is this the Movement of Gold? This force associated with the circular motion seems to be the only constant ... its presence it's always revealed, both in the Micro as at the Macro Cosmos ... very interesting!

However, the orbits of these stars, planets and particles are not exactly circular. Such orbits are elliptical. And why are the orbits elliptical?

Again, apparently, we seem to be doing a simple question ... but I wonder why are the orbits elliptic?!

Knowing that the orbits are elliptical and are in accordance with the laws of Kepler; that the orbits of the planets are ellipses with the Sun occupying one of the focus; it does not explain why these orbits are elliptical.

Why would Nature prefers elliptical orbits rather than perfectly circular. What was the reason that made these orbits elliptical?

If the Gravitational Force and the interaction between the planets and the Sun was only depending of the Newton's law for Gravity; of the Gravitational constant, masses and the distances between the objects involved; then, once we know that the Force of Gravity must be distributed in a uniform way around the objects, producing and equal intensity throughout these stars and around them , distributing it equally ways to both sides just like a sphere shield and in accordance with the Law of the inverse of the square of the distance, shouldn't the orbits of the planets be round and circular?!

Well, but the orbits are elliptical...

Let us imagine if everything had a circular orbit, what consequences would that lead and produce in Nature?

As Gravity acts in accordance with the law of the inverse of the square of the distance, in general we know that the more distant to the source of Gravity the weaker is the gravitational interaction. If the orbit of a planet would have a circular trajectory, this kind of movement would have direct implication in stabilization of the orbit of the planet, putting this one in a critical situation and risk.

A circular trajectory would conditionate the balance of the system, because it would be enough and sufficient to note that the inner part of the planet is always under a greater gravitational pull; the side of the planet closer to the sun and closer to the central source of Gravity; the side of the planet which is far less from the Sun is constantly more attracted and this little attraction is enough to divert the planet a few millimeters of its route. As the gravitational force is already a little stronger in this innermost orbit, the side closer to the center would be even more attracted leading the planet to consecutively inner orbits and so on, that slightly attraction would lead that all the planets, sooner or later, would all fall into spiral towards the Sun!

Not to mention the other external gravitational influences, which would make the balance of the system even more precarious.

Now, let us rephrase the question: Why are the orbits elliptical?

And we can conclude with the following answer: Because Nature is smart.

Nature is very intelligent and knows perfectly well that a circular orbit would not work.

To avoid this balance too much unstable, Nature found a better solution: Nature made the orbits elliptical

An elliptical orbit has its advantages: First, this type of orbit requires that the speed of the planet cannot be constant. When the planet is closer to the Sun it feels the gravitational force of attraction and this makes the earth get acceleration, thus, gains more speed, as such, carries more Kinetic Energy and a greater Inertia.

When the planet is nearest the Sun (perihelion), this is when the velocity vector is perpendicular to the gravitational pull and it reaches the maximum value and the inertia reaches also its maximum value, so the planet continues its route but now to a point farther from the Sun (aphelion). On its path the speed of the planet is decreasing due to the gravitational pull that pulls back, successively reducing the degree of inertia of the planetary traveler, until the speed reaches a minimum and the planet is forced to turn back, powerless, this planet has no more chance to fight the gravitational pull constantly in action, recovering its trajectory closer to the sun and it follows again the same cycle again.

This type of movement is practically constant and eternal, just like the movement of a pendulum, and prevents the collapse of the planets in spirals.

The secret of this movement is reflected in a slight unbalance between two forces: the Centripetal Force (where the gravity pulls the planet for an innermost orbit) and the Centrifugal Force (where inertia requires that the planet moves far away in to a more external orbit).

Let us remember that a grandfather clock the Mechanical energy is preserved and the system is self-sufficient and we do not need to constantly push the pendulum to make the clock work.

This form of movement is almost magical, always constant in time and very precisely. It's like a perpetual motion machine, eternal in time, or almost ... just need 'someone' to give the original rap and the very first ticking.

Now, from the Astrophysics in to Microphysics:

Nature provides many analogies, as if this was not the same Universe that we are considering.

Also within the atom there is movement and acceleration, in fact, there is a lot of acceleration. We can look at the atom as if it was a constellation of particles instead of a constellation of planets. But what allows the balance of this little system, how can we understand the electrodynamics stability of an atom and its constituent particles?

It is known that in this case, we can discard the almost gravitational pull. What keeps the electrons attached together and close to the core is the electromagnetic influence.

In general, we can set the same type of question: if opposite charges attract and if electromagnetic force between protons and electrons is also attractive, how come the electrons do not precipitate into the nucleus?

In order to answer this question, we could adopt the same model of Gravity, and conclude that, within the atoms, the electrons do not remain in circular orbits, and they also move around the nucleus with velocity and acceleration, we could assume that they also found the solution of balance if they describe 'elliptical orbits'.

There is, in fact, a model description of the electrons orbits. The design of these models shows that the orbits of the electrons are three-dimensional ellipsoid. In this case, the 'elliptical orbits' have something more to add ... these 'elliptical orbits' are related with the 'quantum jumps', which is the physical process that moves the electron closer and farther from the nucleus. The energy changing state ... to move from one level to another an electron must perform a jump...

Currently, there are two major issues that most worry the particle physicists:

1- The first, is to know why accelerated electric charges emit radiation only when they make a transition between energy levels, and when they are in their ground state they do not emit electromagnetic radiation;

2- The second, is to understand why these accelerated electric charges in their ground state do not lose all of their energy and radiation.

The first part of question is quite simple to explain, the second, already requires a bit more attention, careful and abstraction.

It appears that the emission of electromagnetic radiation only occurs when an electron moves into a lower energy level. Only in the fundamental energy level is that the atom recovers its natural state of balance and in this specific orbital the electron no longer emits electromagnetic radiation.

So far everything looks right. The problem now consists in understanding why this accelerated electron around the nucleus, in its fundamental state, does not emit any kind of radiation.

Maybe the main problem relies on we continue to search for the wrong type of energy!

Although the atom in its fundamental state does not emit electromagnetic energy, we may not ever forget that this atom continues to emit energy ... delivering another source of energy into space ... Gravitational Energy, or more precisely, Gravitational Radiation!

We could speculate that Quantum Transitions allows emitting Electromagnetic Waves; and that, Quantum Stability emits Gravitational Waves.

This will solve our problem that accelerated charges do continuously emit some kind of radiation. The atom doesn't stop of emitting radiation. The radiation never ceases to be emitted, the only particularity is that the atom emits two forms of radiation: the Electromagnetic Radiation and Gravitational Radiation!

Within this model we could consider the presence of waves of energy in the form of Gravitational Radiation...

Now the second part of question:

Let us picture and imagine an atomic nucleus that needs to be protected from the outside. The shield is made involving the nucleus of an electrical atmosphere with a negative charge, electron particles. This electrical atmosphere protects the core of the constant bombardment of foreign particles, very energetic, as photons.

A photon can come from all directions, that is why the density of negative charge must be very well distributed throughout the atom, with a homogeneity density involving the nucleus. In this way, the electrons are not properly located, in the form of concentrated particles, their density and their electrical charge is distributed almost uniformly expanded around the nucleus. Therefore we can say that the electron is everywhere and anywhere ... that they have the gift of ubiquity!

And what happens when a photon collides with the atom, with this electrical atmosphere?

We can say that at this moment the negative atmosphere absorbs energy; that the energy delivered by the photon is absorbed by an electron.

But what is exactly happening at this precisely moment that the electrical atmosphere absorbs the photons energy?

Now ... the most abstract part of the process: When the photon collides with the electrical atmosphere it materializes the electron at that point and this extra energy allows the electron to achieve a higher energy level and a more distant orbital. We should note that with the absorption of energy it happens the concentration of the electrical charge, the materialization of the electron as a distinct particle, and this process allows the electron to gain freedom and enough energy to repel himself from the nucleus, with this process the radius of the atom is changed, becoming larger.

Note that the absorption of energy does not change the value of the electron charge, it only excites the particle, making it more energetic, because the electric charge, as we know, is a Universal Constant.

If the radius of the atom is changed, but the amount of negative charge is the same (we can still relate the atom as having the exact same number of electrons), but now we can say that the electrical density is more expanded, so the density of charge per unit volume is lower. In the periphery of the atom, in a more distant point, the electrical charge has got more freedom, because it does not feel that much of attraction from the nucleus, but it also has less electrical energy density, once we know that as the radius increases the electrical density is more expanded, and this lower density is insufficient to impulse the electron to leave and abandon the atom, therefore the desertion of the electron is not possible.

This is the point and the exact moment in time that allows the entering of the Electromagnetic Force of attraction between protons and electrons, which begins to be felt. Because, this force also works at the same way as the law of the inverse of the square of the distance and the strength of the proton from inside the nucleus begins to attract the negative charge involving the atom, reducing the radius of the atom into a shorter value, and increasing the density of the negative electrical charge of the atom.

During this process the electrons themselves are powerless to escape the attraction of the nucleus and they surrender! Inevitably weaken they lose the extra energy gained by issuing to external space electromagnetic radiation in the form of photons and they fall down into an inner orbital, closer to the nucleus. And this is when another photon collides with the atom and it begins the same process again ... ad infinitum!

... The photon leaves the atom ... the photon enters the atom; it is energy coming in and it is energy that comes out ... this process is repeated in time in a constant and very precisely way ... just like a pendulum!

Before finalizing our theme of Gravitational Radiation, let us highlight a very common experience, absolutely simple and almost trivial.

For example, let's think about what happens when a piece of metal is gradually heated:

What we are doing is to provide energy to the metal. What happens within the atoms of the metal is that the electrons are absorbing energy and that induces the necessary excitement that causes the electrons to move more rapidly, in a constant stage of more agitation. Since electromagnetic radiation is produced when ever electrical charges are excited, every time an electron is accelerated, the metal begins to transmit electromagnetic radiation in the form of heat radiation, or more precisely, infrared radiation, so the metal achieves a higher temperature and gets heated.

If we continue to heat the metal, before seeing any visible radiation, the metal gets a little bit more heated and the particles begin to vibrate a little faster. With the increase of energy also increases the frequency, until it reaches a point where the metal begins to emit visible light. If we continue to provide energy, the visible light, is initially red, then turn into yellow, then white and finally blue.

So far, nothing new. All seems well...

When we start heating the metal we are causing the agitation of the several electrons constituent of the material, and so they acquire enough energy to produce low-energy quanta with a large wavelength in the form of infrared radiation. Later, with continued heating, we can verify that there is enough energy to start to release some amount of average frequency and average wavelength, thus, it will only be released some radiation of average energy, which corresponds to visible light. And finally, it starts to show up some quanta of high frequency, responsible for the releasing of a shorter wavelength, which corresponds to the ultraviolet radiation.

This would be the logical process of a metal heating. However, although the theory says that there are many electrons capable of producing low-energy quanta and with a long wavelength, what we see and what the experience show is that, that is not the case! In fact, there is very little emission of low energy radiation with long wavelength!

This extraordinary subtlety escapes to any theoretical explanation.

Let us look carefully at the following graphic, which relates the emission of radiation from a heating object:

Color spectrum of Radiation

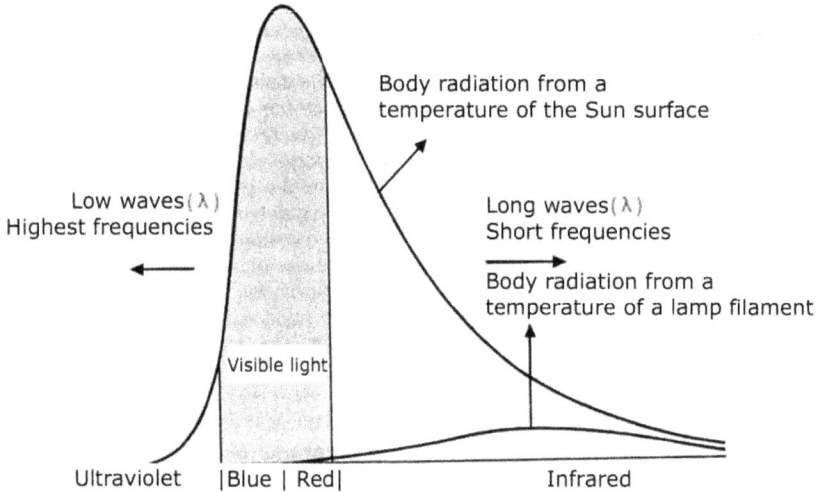

Body radiation from a
temperature of the Sun surface

Low waves(λ)
Highest frequencies

Long waves(λ)
Short frequencies

Body radiation from a
temperature of a lamp filament

Visible light

Ultraviolet |Blue | Red| Infrared

- Radiation emitted by a hot object (Thermal radiation) -

This graph shows the emission of radiation from a heated object. At the right side of the graphic we have the Infrared (long waves) and at the left side we have the Ultraviolet and Visible Light (short waves and high frequency). In this graphic, in which it would be expected to find a large emission of long waves radiation with a very low frequency ...if we look closer at the chart ... we nearly don't find the emission of the long wave radiation! How strange! Doesn't anyone think that this is a little bit strange? It appears that there is some radiation missing! Where did it go all the energy absorbed? Where are the long waves?!

If supposedly there is more electrons with enough energy to emit such a wavelength, so, why aren't they present? Where are the long wavelengths?

It appears that the electrons have a certain difficulty in producing high wavelengths, and consecutively, low frequencies.

According to the Planck constant, the unit of a photon energy is designated as a quantum and this package transports a minimum value of energy: $6,626 \times 10^{-34}$ Js, where the Joule is a measure of energy per second, and this value is very small. There is a minimum quantity of energy. And why is there a minimum energy? Shouldn't the emission of electromagnetic radiation start from value number zero? But no ... there is a minimum quantity of energy. Subtle but curious ... very curious...

What occurs in practice is that the spectrum of emission of electromagnetic waves starts from a minimum value of energy ... which corresponds to the Planck constant.

Below this value, it is not possible to produce Electromagnetic Energy ... but it is possible to produce another kind of energy, right below the Planck scale, an energy whose value is the order of 10^{-38} or 10^{-40} ... that energy is the Gravitational Energy or Gravitational Radiation!

The long wavelengths, the low frequencies are not in the form of electromagnetic radiation ... they are in the form of Gravitational Radiation!

These are the small subtle messages that Nature provides us, giving us the possibility to see and notice the little details. And that is why Gravity is the weakest of all forces!

In Nature, there is no Law of Casualty ... personally I do not believe that there is undefinition and indetermination in Nature or laws of fortune. It seems that everything happens as a result of a cause-effect relationship of previous events, and everything seems to be intentionally organized with a sense and practical application.

The Physics takes us to conclusions and thoughts ever more interesting and surprising, which gives Nature a huge potential of functional intelligence!

Everything works perfectly...

Chapter XV

Quantum Theory of Gravity

1. REDEFINITION OF MASS AND GRAVITY

"The most beautiful experience is
the encounter with the unknown."
- Albert Einstein -

Mass! Even the particle physicists don't measure the mass of an electron ... let's say ... with a weighing machine, that is, in kilograms, or Newtons, or any of the units of mass or weight. For example, it is establish that the mass of the electron at rest is 0,511 MeV, this means, millions of electron-volt, which is a measure of amount of energy!

The masses of sub-atomic particles are thus expressed in units of MeV / c^2, usually shortened to MeV. And it is said, for example, that the mass of a proton is 939 MeV. This is the amount of energy that would be produced and released in case the mass of the proton would be destroyed and completely eliminated.

According to Einstein's equation, the energy released by a tiny amount of mass would be huge: $E = m.c^2$

Looking at this equation we can see that it is sufficient to multiply a minimal amount of mass by c^2 to have a result of a huge amount of energy.

But Mass still keeps many secrets for the scientists that, repeatedly, keep trying to reveal the mechanism by which the small particles of atoms are compose with Mass. Still continue to search and try to discover what is the minimum unit of matter, or what is the mechanism that provides the quality of Mass. Because it has always been said that Mass is an intrinsic property of Matter. All of these considerations are very interesting... but, after all, what is Mass?

Indeed, atoms and matter are consisted of electrical charged particles, as such, shouldn't they be considered at least as a portion of the electromagnetic spectrum?!

I even dare to say that it would be desirable, and even more, to give preference to a theory that make up the Gravitation Force and the Electromagnetic Force as having the same source! Different manifestations of the same underlying phenomenon!

Once assumed that the Force of Gravity is a Force of Radiation, perhaps it is now easier to rearrange a new non standard theory, once we have release of a large preconcept and prejudice!

We could start our investigation with the following question: What is the property that all particles with mass have in common?

Let us start by remembering all of the stable particles with mass that we know:

Starting with its identification we have: protons, neutrons, electrons... Quarks some bosons...

What is the common variable in all these particles?

Looking closely, however strange it may seem, all of these particles have a common variable ... the property that these entire particles share is ... electrical charge!!

All particles endowed with mass have to have charge in its constitution?! Curious ... this could be our 1st clue...

The protons are positively charged, the electrons have negative charge, and neutrons are neutral since this resulting charge is made of quarks with fractional charge. The atom itself is also composed with different types of electrical particles. The addition of all of these particles makes the contributions for the electrical charge of the atom, with a total result of a neutral charge. The atoms have neutral electrical charge. And the atoms have mass.

We can consider that the photons and neutrinos have no charge in its constitution, therefore, these particles have no mass.

It seems that mass cannot exist without the presence of electrical charge ... that seems a little bit strange! Is this the correct variable? ... Or maybe not! ... But it is a good evidence.

Is it here any chance of relationship and unification between Electromagnetic Fields and Gravitational Fields? Considering the hypothesis that Gravity is a force of radiation, in what way does that provide gravitational attraction between two objects, and in what way does that give and extent the quality of mass to matter? And what kind of relationship can be established between the property of mass and the property of charge in the middle of this entire scenario? ... Very confusing...

Well, it might be better if we start with a subject at a time ...

But, looking closer to what has been happening over the last three decades, the unification of Gravity with the theory of Electromagnetism has demonstrated to be very inconsistent and so hard and difficult to achieve, so it is more likely that one of these two theories may not be quite correct as we think!

Again, let us star from the beginning:

I stop in this equation, or Newton's law of gravitational force between two masses:

$$F_g = G. \, m.m \, / \, r^2$$

And after this, we can stop in this equation, the Law of Coulomb for electrical force between two charges:

$$F_e = K. \, Q.Q \, / \, r^2$$

As you can see, until here, we have no secret!

Both of these theories describe with success two distinct theories and apparently, unrelated phenomena.

By choosing one of the equations, we could start by checking the veracity of Newton's Law.

Let us start then with our rigorous analysis.

This part of the equation of the Law of Gravity: $m.m/ \, r^2$, we have already seen earlier in this work that it is incorrect, since it leads to an indeterminacy of the infinite problem... because when the radius is equal to zero the force of Gravity becomes infinite. And if the Force of Gravity becomes infinite in the center of every object, everything would collapse...

$$r = 0 \rightarrow F_g = \infty$$

And the other part? The Gravitational constant 'G' ... what is the meaning of G? It is said that it is a constant ... a gravitational constant ... but constant of what? ... Constant that relates masses ... but what properties, specifically, among the masses?

Theoretically, the description of this constant is defined as follows:

The constant of proportionality G is a universal constant of Nature, describing the intensity and proportion of the force with which two masses are mutually attracted.

And that this constant takes the same value for all substances, whatever the composition of these chemical elements, this means that this constant is independent of their constituent elements, of density, weight ... the very own constitution of mass or of matter!

A constant that describes the interaction between two masses is independent of mass itself? Curious...

But we can also address it in another way, describing this constant in another form: that this is a constant between the forces of attraction...

Ah! Yes, now this is more like it and this is a very different sentence.

Based on the observations of Galileo, it found that the acceleration of bodies in free fall is independent of its mass. This means that, disregarding the force of friction, the bodies may have different masses and different weights, however, both bodies fall down at the same speed and touch the ground at the same time, because the acceleration is the only constant ... very interesting.

The value of gravitational constant is $G = 6.6742 \times 10^{-11}$ N.m^2/Kg^2 or also in their official notation $G = 6.6726 \times 10^{-11}$ m^3/Kg.s^2 (these units refer to the cube of the distance, divided by the mass multiplied by the square of the time) and was obtained experimentally with a device designed by Sir Henry Cavendish in 1798, the torsion balance.

The procedure to this measure was the following: Put up two objects, or two spherical masses, suspended by a thread but united as a dumbbell made of a very light rod. Then, place it down to the bottom another fixed dumbbell consisting of heavier spheres and with higher volume.

After that, we try to measure what is the gravitational attraction that occurs when approaching the spheres. The wire suspending the small ball is forced to twist due to the gravitational pull caused by the heavier spheres, making an angle with respect to the axis of origin. The extent of this small angle of rotation is measured and related to the force of attraction.

This experience, in practice, requires a few more details, however, the basic idea is the one presented and the conclusion is the following:

That the gravitational force of attraction is relatively weak and that, for example, in practice if we considered two spheres of one kilogram each, placed at a distance of one meter in relation to its center of gravity, we will detect a force of attraction between these two bodies of $6,67 \times 10^{-11}$ Newton.

In the Newton's law we can simply replace m = 1 and r = 1 in the equation of gravity to obtain the value of the force measured. It is postulate a minimum value for the gravitational attraction and this minimum value is placed as G in the Gravity equation and designated as a universal constant... which is in fact a value of a force!

A minimum strength, a very small force of attraction, but it is the value resulting from the force of attraction between two masses with one kilogram, separated by a distance of one meter. I do not think we can consider this as a Universal Constant ... at least, in my point of view, this constant should need some reviewing and have a better explanation…

But, some think that this relation is useful and that will fit in the Gravity equation and resolve many problems, and so there it is, we introduce G in the Newton's equation, and everything works perfectly ... very convenient!

But, in fact, it seems that this constant it is not working that well. What we see in practice is that the Gravitational Constant G is rather difficult to determine.

The oldest constant of Physics, the Universal constant of Gravity, is in fact the most difficult to determinate with precision and has proved by far to be the most inconsistently constant to define with good accuracy.

Usually, all other universal physical constants can be measured with a precision that is up to eight decimal places, or even more, for G, the differences appear after the third decimal place, sometimes even before!

The error in the measurement of G is so big that it is too high to be used in studies of gravity and space operations. To avoid that error we take as a reference another celestial body with a mass 'm' higher, and with that, in practice, what we do obtain is a new G!

The experimental results do not match and so we think that the main problem is on the measure devices. Then comes up some new investigators with new ideas for new devices, more modern and more sophisticated, they go into the field and adventure into the measurement of G and find out, once again, what they will get as a result is always a different value … which is very annoying!

This reminds me the inverse story of 'c'. The early steps for the determination of this constant was made in the direction to obtain different values for the speed of light, but insistently, the value of 'c' always wanted to remain constant. Now we are seeing the opposite, constantly trying and demanding the same value for G, but at every measure different values are always found!

This number of the gravitational constant insists to be inconsistent and inaccurate, and the truth is that until today no one knows its exact value!

It is generally believed that the problem is in the measuring devices which cannot measure this constant with quite precision and perfect accuracy.

I wonder why the oldest constant of Physics it is so enigmatic…

Furthermore, we could try another approach and try to accept and assume the evidence of facts. And what are the evidences?

I leave you to reflect for a moment.

Before going any further, let us see how with a few small experiments and some simple calculations, we can explain and clarify a little bit more some of the anomalies of our Gravitational Constant.

A work entitled *"Geophysical evidence for non-Newtonian Gravity"* published in 1981 by F. D. Stacey and G. J.Tuck, has developed some measurements of G below sea level, at the bottom of mine.

What was found in these measurements was that amazingly the gravitational constant G had achieves values up to 1% above the official, much superior than the measurements that are performed in the laboratory at the surface of the Earth. And, the greater the depth, the greater was the value found for G!

Shouldn't G be always constant?!

What this means is that the force of attraction between two spheres it is no longer the same, is different. Thus, the force of Newton will be higher as it increases the depth ... for the same spheres at the same distance...!

If the mass is a measurement of the number of atoms of the spheres and if the amount of mass remains unchanged, both the spheres and the planet Earth, why is the pattern of gravitational attraction different?

The same experience of Cavendish leads to different results! What is the variable in this experiment? It is not the mass for sure.

Are we still so sure that the gravitational attraction is a function of masses?

Another study published in 1924 by Charles F. Brush, called: *"Some new experiments in Gravitation"*, shows us photographic pictures which demonstrates that heavier bodies made of metal atoms and more dense tend to have more force of gravitational attraction and fall faster into the floor than bodies with the same mass, but less dense or with lower atomic number. This difference is minimal but measurable.

Why does this happen? Another anomaly of G?!

This new anomaly leads us to introduce again the following observation: that the amount of electrical charge, number of the electrons constituent of the atom has an influence on the amount of mass, or the amount of gravitational field which is produced. But how is this possible?

Finally, the most enigmatic experience of all, which completely challenges the incontestable validity of the Law of Gravity.

In 1798, Henry Cavendish had the curiosity to pursue the experience of the torsion balance but in a slightly different way. While he measures the gravitational constant he has resolved to heat, with fire, both spheres.

Surprisingly, he had found that the force of attraction between the two spheres had increased considerably, therefore he has determined a value for G much higher!

This experience has been challenging the Classical Physics for over 200 years!! And all efforts made to try to explain this phenomenon have been in vain or very weak.

After all, what is generating more Gravity? Again, I will repeat ... it is not the mass, for sure!

Advances in Science are not always made forward, sometimes it seems that we are moving backwards...

Now, let us repeat the question: what is the evidence of these facts?

The evidence is, at least, that G is a constant which does not seem very constant at all ... and with this kind of evidence we could think that perhaps G is not even a Universal Constant!

This could be our 2nd evidence: the force of Gravity that relies so much in the gravitational constant G, the universal constant of Newton's formula, this is not even a constant but a variable, a parameter of location, which can take values and results always different, depending on the local and external conditions in which the measurement is being determinated.

We should remind that Universal Constants are references of the Cosmos. As such, a Universal Constant does not change, does not vary, it is universal because it applies to the entire Universe. And it is certainly independent of external conditions and locations of measurement.

The experimental evidence that G varies, it shows that this is not an innate or a fundamental property of mass, and it is a proof that the gravitational constant G is certainly not a Universal Constant at all!

Without wishing to complicate Physics, we must accept the facts! As much they seem to contradict our fundamental ideas...

However, assuming that there is no gravitational constant with universal application, there still is an apparent constant of attraction with application for a specific location.

Probably the approximate value of G that appears in our equation is a consequence, a relationship between other properties inherent to the masses. All we can ask at this moment is what could be those other properties?!

Before we continue, I will leave you another comment about the Law of Gravity, which is:

Imagine an astronaut, for example. We know that an astronaut on the moon does not have the same weight as if he were on the Earth, nor have the same weight as in Jupiter. Because in Jupiter the extreme gravitational forces tend to compress the mater and then an astronaut would feel his body to weigh increasingly, becoming very heavy and wanting to collapse on itself. Its weight would have a value almost three times bigger that of Earth, a weight impossible to support, a weight which no human body can stand.

It is good to know we are aware that we cannot send man expeditions to Jupiter. However, we have already sent astronauts to the Moon, and with these conditions we might assume the opposite. We know that an astronaut on the moon weighs significantly less one-sixth that on Earth, and therefore it is said that the reason why this happens is because the astronaut is less exposed to gravitational influences, which is why his weight is lower in that place. However, despite being less subject to gravitational influences, his body does not extend, doesn't expand, so it does not change the form! The geometry of the body remains ... the geometry of the body remains the same...

What I have said may seem an irrational idea, because we know that atoms are minimally stable, independent of the intensity of Gravity. This implies that the internal forces of the atoms tend to adapt to each location, so that the micro atomic system remains stable. The stability of an atom depends directly of balance of the internal electromagnetic forces.

Despite the influence that inter-atomic binding energies of molecular and covalent connections, that probably will take more effort to maintain stability and balance of the atomic system, since tolerating the absence of gravity must require a great effort by the body of the astronaut, we must consider that there is indeed a factor that has been changed: the gravitational intensity.

In practice, we can resume these observations as the following:

1st The amount of mass: = is maintained => Because the mass is a measure of the number of atoms, and it remains the same number of atoms;

2nd The value of Weight: = it changes => The weight changes considerably, because the body is exposed to a lower 'gravitational' field;

3rd The Form: = remains the same => Because it appears that the geometry and gravitational attraction of the body does not change considerably.

Again: the gravitational attraction of the astronauts body does not change significantly but the weight changes a lot … very subtle ... strange ... but curious!

This leads us to suppose that the gravitational attraction is quite independent of weight, since the weight change is substantially but not the form. So we could ask again the same question: what is Gravity?

We believe that this force is responsible for transmitting weight and also the form of an object. But as we can see, these two properties of Mass do not appear to be directly related...

I do not wish to confuse the concepts, but it seems to me that there is something under here that wishes to be very well disguised!

Are we still so sure that weight and gravitational attraction are one and the same thing?

The Law of Gravity is becoming each further more complicated!

Soon we will see a light at the end of the tunnel...

If Gravity is a force related to Masses, then this force must be very special because it does not allow many analogies. For example, when a gas is subjected to enormous forces of pressure, it tends to compress; making an analogy, we can consider that if the gas is exposed to a lower pressure, tends to expand. The forces that are acting on the gas vary, so the gas changes its shape. Right?

But in our case, Gravity does not change the form, even when the forces that are acting on the masses are suppress or have substantial differences! You may not agree, but I find this a little bit strange! Quite strange! What kind of force is this?

Even though it's of common knowledge that Gravity is considered as a Field Force, it must be noted, inevitably, a 3rd evidence: that this Force of Gravity it is not, unquestionably, a Mechanical Force or a contact force.

What we usually call of Force of Gravity cannot be a universal property of Mass. This force linked to the attraction is not proportional to the amount of mass of each substance, but it is related to another property of matter!! ...

Without wishing to cause any inconvenient, and with all the respect and admiration that I share for Sir Isaac Newton, I would say that the Theory of Gravity is losing points ... many points. So, it would be good and preferable if we keep as a reference the Theory of Electromagnetism.

This is the moment which seems that I will have to say something completely absurd. Listen carefully: If Gravity is not a mechanical force, what kind of force can it be? Again, there are not many hypothesis left ... it can only be a field force ... a force of radiation ... an electromagnetic force!

Try to follow my thoughts and again, without any preconcept or prejudice.

If Gravity is not synonymous of mass not even of weight; then, the only thing that we can say with some certainty is that Gravity is a force of attraction between atoms.

Everyone agrees?

Very well, considering Gravity as a force of attraction, what other forces of attraction do we know in Nature?

Let us look at the magnetism. This phenomenon known for centuries shows us how two magnets can attract so quickly due to a mysterious force that unites them. Is there any force more attractive than this? What is this magnetism?

Magnetic fields are everywhere, naturally produced and artificially produced. The largest natural magnetic field that involves us is the one created by our planet Earth, the terrestrial magnetic field. Other manifestations of magnetic attraction are present in small magnets.

Magnetic fields can also be produced artificially, every time an electrical device is in operation. But man can not directly feel this magnetism, it is an invisible force that works quietly through the empty space without being notest...

The magnetic effects are a by-product, a secondary manifestation of a fundamental force, which results of the behavior between particles with electrical charges.

All particles containing electrical charges in motion create electromagnetic fields. Nature does not have isolated magnetic charges. The magnetism is a consequence of the movement of electrical charges which can be produced by electrons, protons, atoms, planets or stars and all of these objects can create magnetic fields.

All particles have their own magnetic field, resulting from its rotation, however, it may also arise another additional field, if the particle has a speed of translation. Wherever there is movement, there is magnetic field.

Even neutrons have their own magnetic field. Although these particles are electrically neutral, they do not cease to consist of quarks with fractional charge, as such, they also create magnetic fields.

We can conclude that all atoms have their own magnetic field and the intensity of this field also decreases according to the law of the inverse of the square of the distance, as stated in the theory of electromagnetism.

One of the properties of objects in rotation is called angular moment, which is a measure of the amount of rotation. A particularity of the angular moment is that this force can be transferred.

For example, if we are sitting still in a rotating bank while we are holding a bicycle wheel horizontally, so that its axis remains upright, if we put the bicycle wheel to rotate and begin to reverse the direction of its axis, surprisingly, we will also start to rotate in our rotating chair. The reason why this happens is because the angular moment can be transferred ... very interesting!

Back to the atom. In this case we have particles that are in a constant movement of rotation and therefore they can also create an intrinsic angular moment, however, since these particles have internal charges in its constitution, in addition to generating an angular moment these particles can also generate a magnetic moment.

Since the magnetic moment is a vectorial quantity, and assuming that all the magnetic vectors are always aligned, we can speculate that the magnetic moment of an atom is the resulting of the vectorial sum of all these small contributions, so that the moment or the magnetic field produced by an atom is a constant in time.

It is the intrinsic magnetic moments of the particles that are responsible for the macroscopic effects of magnetism. The magnetic moment of a system is a measure and a consequence of the intensity of the magnetic source. It is a quantification of the system's internal magnetism. This source is generated by magnetic particles constituents of the atom, the same number of protons neutrons and electrons, whose ratio is always proportional and constant, unless exceptions.

Each atom creates a magnetic field, each group of atoms creates a magnetic field slightly larger, and larger clusters of atoms can create bigger magnetic fields.

The existence of this tiny and subtle magnetic field that each atom produces is simply a magnetic attraction, with the same behavior as a very small magnet.

Unfortunately, the magnetic field generated by an atom is very weak. According to the theory of electromagnetism its magnitude it's always weaker with further distance, thus we can consider that the magnetic field produced by an atom is almost negligible and as such we cannot suppose that this force is responsible for a universal attraction of matter, for the Law of Gravity.

It is true that this field is extremely weak to achieve and contact with another distant atom and be able to attract him ... but maybe there still missing another variable in this model that makes all the difference...

It is now that we should consider another phenomenon: the electromagnetism manifestation at the macrocosms. As we know, the electron-photon interaction produces the Electromagnetism. This electromagnetic radiation is produced to outside the atom, for the macrocosm. The emission of this radiation is then spread throughout the space in the form of electromagnetic waves at the speed of light.

The theory that I wish to demonstrate is the following: The emission of the electromagnetic field does not travel alone. Along with this is associated another field, the magnetic field of the atom. What I think is that, in some way, the magnetic moment of the small atom is transferred trough the way of electromagnetism emission to the macrocosm, and this force of attraction, the magnetic momentum of the atom, although small, can achieve infinite distances.

If in Classical Physics, the angular kinetic moment is transferred, much probably, the angular magnetic moment can also be transferred.

And this would be the phenomenon responsible for causing Gravity!

A simple experiment shows us that the chemical unfolding of the Hydrogen spectrum has two very distinct fine lines, rather than a single line. What is the meaning of these lines? The physicist response is still somehow inconclusive, however, we could say that this flaw in the spectrum is related to the magnetic moment of the electron or of the atom itself. We could suppose that these two lines of Hydrogen spectrum show the absorption of two different sources of radiation: Electromagnetic Radiation and Gravitational Radiation.

These two lines may clearly demonstrate the existence of the absorption and emission of a double radiation resulting from an internal magnetic field (gravitational source) and an external electromagnetic field (classical electromagnetic source), exchanged and circulated continuously throughout the space at speed of light.

The emission of this new gravitational field obeys always to a constant value, since each atom has got an exact proportion of its constituent fundamental particles, the same number of electrons, protons and neutrons. And with this we can conclude that the amount of charge is, in general, always constant and equal. The ratio of magnetic moment is therefore a constant ... very interesting...

The advantage of this New Theory of Gravity is that the constant of attraction does not need necessarily to have a fixed value, the gravitational constant does not have to be always constant, it can be variable.

Take the example of the production of artificial small magnets. These industrial magnets can be made out of a diamagnetic material expose to an external and intense electromagnetic field. Depending of the intensity of the field, you get a magnet with more or less magnetization. That is, if you wish to have a stronger magnet, more magnetic, you can simply submit the material to a stronger electromagnetic field. And we know that an electromagnetic field is more intense as more rapid it changes.

Now, back to the macrocosm. If we consider a planet or a star in which its speed of rotation is higher, this will be reflected in the production of a stronger external electromagnetic field.

The subtlety of Gravity is as follows:

If the star exposes itself to a stronger and more intense electromagnetic field, it will happen the same analogy as our artificial small magnets, which is: the star intensifies the inner magnetic field of the system, the residual internal magnetic momentum of each atom increases, which in practice it will be reflected in an increase of the gravitational constant!

The same applies to the experience of Cavendish. While we are heating the spheres we are transferring kinetic energy to electrons, and as a consequence these particles move more quickly, shacking rapidly, and this increases and intensify the external electromagnetic field which is produced. If the spheres involves itself to a stronger electromagnetic field this will also interfere with the internal magnetic momentum of each particle, therefore, this residual internal magnetic momentum, the gravitational   field of attraction, also increases and inevitably the G 'constant' must achieve a higher value!!

The inconstant G does not reveal a surprise … actually it shows the way Nature works…

In addition to this theory we must say that the magnetization is a phenomenon that is only possible because it produces the alignment of the domain.

In our case, what could be those alignment fields of the particles of Nature?

There is a strange property of matter that has passed by a bit unnoticed, because in fact, nobody knows very well for what it serves: the property that I am referring to is the enigmatic and mysterious Spin.

Why would Nature need to create this property? All particles have a very well defined spin, and this is a fundamental characteristic of matter and it cannot be changed. The spin, is related to the direction of the axis of rotation, or with the intrinsic angular moment of a particle, and is one of the most intriguing properties of quantum physics. Until this moment there is no convincing physical interpretation to explain this particular characteristic of Nature...

Interestingly, all the entire family of fermions, which includes the stable and unstable matter, has got the exact same spin. For us, we should note, particularly, the stable part of this family, the common particles around us.

Quite amazingly, nobody noticed that all particles with stable mass have all an identical property: Spin ½.

The electrons have spin ½; The quarks have spin ½, which are the constituents of the nucleus, therefore, also the protons and neutrons have spin ½, so all atoms have spin ½.

This identical spin to all of these particles means that they all operate in the same direction around the same virtual universal axis. All particles of matter perform a rotation in the same way so that the spin associated with this movement always corresponds to the module 1/2 ... And this is how it is performed the alignment of the domain!

Perfect! The magnetic fields of all these different and distinct particles are related with one universal alignment according to their spin, which is always equal to the module ½, and behave themselves as small magnets spread around the Universe with two poles, the only change that can be made is with a spin of 1/2 or -1/2, to allow the combination of the two poles, let's say, add a North Pole with a South Pole.

The resulting magnetic moment is the total sum of all these individual contributions of each particle, which is a constant ratio of the magnetic moment per unit volume. So that, the final magnetic moment is always constant and always proportional to the amount of mass and number of particles that produces this magnetism, therefore, the constant G arises as a constant of magnetic attraction.

The Spin is the secret of Gravity!!

This revelation is very far from being glorious as it still is incomplete, because the Force of Gravity still has got many more mysteries and secrets to reveal.

If the Gravitational Force is a magnetic force, responsible for the form of the objects, so then what is weight?

The weight as we shall see, it is another story...

The Force of Gravity is responsible for the geometry of the bodies, responsible for the attraction of particles of matter, however it is not directly related to mass or weight. Indeed, these are three distinct concepts, mass, weight, gravitational attraction. As we will see, these qualities are absolutely different and independent!

I do consider these three concepts as being absolutely independent, and I will explain why:

First, the attraction of a body is therefore a function of the attractions of its atoms. However, this attraction it is not a universal and essential property of the body, or of their particles, or of mass. Indeed, it is not even a specific and innate property, but a consequence.

It seems that without an electrical atmosphere we would not have the generation of Gravity, since, according to this theory, it is due to electron-photon interaction that produces the phenomenon of electromagnetism to the outside of the atom and therefore the emission of Gravity into the macrocosm... and this feature is very interesting !

As we can see in Nature, we rarely find solitary atoms. Nature tends to form groups and acquire more complex forms. However, what allows the first approximation of these atoms, it is a force of attraction ... a magnetic force ... the one that we usually call of Gravity Force.

This Magnetic Gravitational Field, appears as a small interference, a slight imbalance in the position of particles, requiring that the matter can no longer stay in a static equilibrium and are forced to move into a state of dynamic equilibrium. The mass has a tendency to agglomerate, but it does not disintegrated, it doesn't collapses on itself, there is no infinite gravitational forces in the center of the nucleus, in the center of mass where the radius is equal to zero, because Gravity does not emerge from the center ... this is the action of the New Force of Gravity!

It is only later that there is the union of atoms. Without wishing to enter in the fields of Chemistry, the union of these atoms is reflected by molecular links, ionic and covalent, much more resistant. The electronic connections of the valence electrons, those which are at a more distant orbital, are very strong and it ensures the construction of chemical elements in a more complex and stable structures such as molecules and crystals. Later, this structure takes a strict and defined form, resulting in the classification of this substance as a solid, a liquid or a gas.

However, the existence of atoms and molecules does not establishes that there must exist a predetermined weight, because the weight is always a variable.

The next big question is try to explain what is the mechanism that gives mass and weight to matter.

In a very short analogy, very succinct and without any mysteries, I would say that if the photons are responsible for transferring the electrical charge, then, our neutrinos or gravitons are responsible for transferring the mass.

We cannot forget that these are the fundamental properties of the Cosmos: Charge and Mass, two very exotic properties!

The neutrinos would be the particles responsible for building and produce all the mass that we are made of!

Until now, our neutrinos have little affected the lives of the particle physicists. It would be good if we could give a little more attention to these subtle particles and develop further studies.

The existence of the neutrino was first postulated by the theoretical physicist Wolfgang Pauli in 1931. Pauli based this hypothesis in apparent non conservation of energy and time in a certain specifically radioactive declines, the Beta disintegration. This type of neutron disintegration resulted in the appearance of two new particles, the proton and electron, but apparently there was a tiny quantity of energy missing. Pauli suggested that the missing energy would be carried by a neutral and invisible particle. Later, Enrico Fermi baptized the name of this new particle as a neutrino. In 1959 it was finally discovered a new particle which corresponded exactly to the characteristics of the neutrino.

The neutrinos are neutral elementary particles that interact with matter only through the Weak Nuclear Force. However, the process of production of these neutrinos can be provided from very diverse sources.

Most of the neutrinos production happens in our Sun. These small particles are continuously generated in nuclear reactions inside the Sun and in other stars. The neutrinos are the most important component of the whole cosmic radiation that constantly arrives at our little planet Earth.

The most fascinating feature of this small particle is that neutrinos barely interact with matter, since this particle does not have electric charge and probably does not have any mass, this ghost particle can cross the planet very easily, without reacting with matter, because for these neutrinos the whole matter is almost transparent, and they can get through it without any difficulty!

The solar neutrinos arrive from all directions at every moment, coming across the space and through the planet Earth, extending until the ends of the Universe.

These neutrinos may have different distributions of energy, depending on the nuclear reaction that produced them. Given the luminosity of the Sun, it is possible to calculate the number of neutrinos generated at every second. If two neutrinos are emitted per 28 million eV, as they expand in all directions across the spherical area of the solar surface, it is estimated that the number of neutrinos that reaches the surface of planet Earth is, approximately, 60 billion of neutrinos per cm^2 per second!

And as bigger the mass of the star, the higher is the amount of neutrinos that are generated. It is estimated that currently there are 10 billion neutrinos per proton. These particles are practically undetectable and probably, at least, as abundant as the photons. And at the same way as photons these invisible particles must travel at the speed of light.

Still, the production of these particles is not limited to nuclear reactions of stars. The neutrinos can also be produced inside the Earth, through the radioactivity disintegration of some elements; in nuclear power centrals installed at the surface of our planet; and even by the human being, as a result of specific reactions with Potassium atoms that compose our bodies.

The truth is that, a human body produces 20 million neutrinos per hour; besides that, it is crossed by 100 billion neutrinos coming from nuclear centrals; and finally it is crossed by another 50 trillion neutrinos coming from the Sun!

Therefore, it is not an absurd to say that we are crossed by trillions of neutrinos in a short space of time!

Amazing ... neutrinos get trough us at every moment without being notice ...

These particles, so subtle and omnipresent, were created almost since the beginning of the Universe, the entire evolution of Nature has had to rely on these structures, therefore, these particles must have a fundamental function ...

It is an idea a little difficult to prove, the materialization of matter through the neutrinos, but only this way makes any sense, other wise ... why would Nature create so many neutrinos? ... Take away all the neutrinos of the Universe and everything will vanish into dust and dilute into air ... literally!

The neutrinos would be the transmitters of the material energy, a property that gives mass to matter, as the photons are the carriers of the electromagnetic energy, a property which gives electrical charge to matter. These would be the two essential properties of atomic elements, electrical charge and mass, mediated by these two particles: photons and neutrinos.

Thus, according to this analysis, it seems that we should reconsider and divide the structure of our Old Force of Gravity into two components:

1st Component - The Magnetic Gravitational Force, responsible for the geometric shape and form;

2nd Component - The Material Gravitational Force, responsible for transmitting the quality of mass.

These two forces combined together and related, gives us the illusion of the existence of a single force, because they fit almost perfectly in order to compose a Secondary Force that we normally call of Gravity Force!

The sad conclusion is, that there is no Force of Gravity as an endemic and original force of the Cosmos, this exotic force deludes us with its beauty as a magnificent hybrid!

With this information, we are forced to readjust the major structural forces of the Universe, reconsidering for them on a new model, a new Equation of the Cosmos:

Weak Force = Strong Force + Electro-Magnetic-Gravitational Force + Material Force

Nevertheless, it would be necessary to postulate a Theory for Neutrinos, prepare new experiences, in order to better understand what is their real interaction with matter.

Recent experience in this area, that wants to count the number of neutrinos that constantly crosses our planet Earth, have found that the number of neutrinos coming out is not exactly equal to the number of neutrinos coming in. There is, therefore, a deficit of neutrinos.

There are some suggestions that have been made in order to provide an explanation for this phenomenon. Some scientists believe that the

127

equipment and the way some experiences are made turns out impossible to count certain species of neutrinos; other scientists believe that there is a transformation of this particle into another ... but we can venture another alternative which is: the neutrinos that are passing through our planet are absorbed by atoms, interfere and mixed with the matter.

It would be important to inquiry and investigate whether these neutrinos actually mix with the matter, thus confirming that they are responsible for transmitting the quality of mass that we are made.

Another interesting property of this particle is the Spin, again. The neutrinos also have spin ½. Interestingly, this mediator of the material interaction does not have an integer spin as the other mediators of the other forces of Nature. The photon, for example, mediator of electromagnetic force has a full spin 1.

The reason for this spin could be related to another fundamental characteristic of matter, the fermions ... all particles with mass have spin ½. The ability that neutrinos have to overcome the electronic atmosphere without disturbing the atom, once this particle has no charge at its constitution therefore does not interfere with the electrons, allowing it to achieve the core and get inside the nucleus without any difficulty. The fact of having spin ½ is the password for entry. The neutrinos would be the particles responsible for transmitting throughout space the same momentum and material energy.

Relate this property with the variation of weight consists in another challenge. We know that an exact same mass can have different weights. But where does it come from, what's the origin of this strength of the weight?

If the weight is not an innate characteristic of the masses and is not directly related to the magnetic force of gravity, then, what is this 'weight ', which is beginning to weigh on our thinking?

To clearly understand this phenomenon I would say that the weight is closely linked with two unique properties raised by Einstein in his Principle of Equivalence and already studied and presented before by Galileo. The abstract characteristics which I am referring to are the inertia and acceleration. These characteristics are deeply related to the concept of weight.

To recapitulate and summarize it is suggested the following:

1. Gravitational Attraction => It is a Magnetic force;

2. Mass => It is a form of energy, is a Material force;

3. Weight => It is a force of acceleration caused by the space-time deformation as a result of the presence of a large amount of energy. The mass, or the material-energy present causes the material deformation of the space tissue, it changes the geometric form of space; and depending on the value of the material-energy present, more or less curvature will be presented in the space tissue. Under these conditions, every single mass is required and forced to perform an acceleration and win a state of inertia, from which derives the status of weight.

And that is why it establishes a principle of equivalence between:

Inertial mass = Gravitational mass

Gravity = Acceleration

These three forces combined together overlap at the same time, making us think that this is a single force, which erroneously is called as Force of Gravity. But as you can see it is not only one force that we must considered, but in fact they are three!

Since these three forces are consecutive consequences of each other, in practice it is almost impossible to separate them and distinguish them from which, and somehow, it does not arise any problem in maintaining our tribute to Newton and continue to designate it by Gravity Force. Without, however, never forget that gravitational attraction, mass and weight are quite different concepts.

We will now review the state and point of situation of our astronaut.

Forgive now the flow of my words, because this description actually requires a higher degree of abstraction.

The 'Gravity' from the astronaut operates through three components:

At first place, we must consider that the gravitational attraction is almost constant, therefore the geometric shape of the astronaut practically does not change, the astronaut does not change its form, because the gravitational attraction is a magnetic attraction and a radiation force, thus, if it is not a mechanical or a contact force the appearance of the astronaut does not expand and it is maintained;

Second, even having an internal reserve of own production of neutrinos that ensures, guaranties and gives its structure material, the astronaut is exposing himself to a lower flow of these particles, because it is farther from Earth and the production of neutrinos generated by the lunar satellite it is considerably lower, since this satellite has ceased almost all of its internal geological activity.

Third, as a result of these conditions, although the amount of mass 'm' remains exactly the same, the material energy surrounding the

THE TRAVEL IN TIME

astronaut is clearly reduced. And if the material energy that is affecting the structure of space-time is lower, this will be reflected in the curvature of space which will be less marked, thus, if the declivity is not much accentuated so the acceleration 'a' of the astronaut is smaller, and therefore the astronaut presents less weight!

And the weight 'W' is represented just the same old classical formula:

$$W = m.a$$

Because Weight is a force that arises as a result of the material energy acceleration. Its manifestation is a direct consequence of changes and distortions of the space-time which a mass is involved.

This is the first theory that explains and justifies the fact that there is a direct relationship between these concepts: Gravity, Acceleration and Inertia.

Explains what happens and why it happens, without having to establish any confrontation between Newton's and Einstein's Gravity Theories, because both theories have correct features.

Gravity is not only a geometric force representing the deformation of space-time. In Gravity there is mass-energy present and there is also deformation of space-time.

With that said, perhaps it is not necessary to use and appeal for new concepts, new fields, new particles ... including the greatest Boson of Higgs ...

And, talking about Higgs particle ... the reason why it is being building the LHC in Geneva, the Large Hadron Collider ... between the enigmatic Graviton and the grandiose Higgs, another question arises immediately in my mind: The Quantization of Matter...

Chapter XVI

Quantum Theory of Gravity

2. QUANTIZATION OF MATTER

"The atoms are not things."
- Werner Heisenberg -

In general, in Physics, a field is always associated with a particle and the existence of Higgs Boson would explain why all bodies have a tendency to resist a change of speed. That is, the fact that all bodies have inertia is seen, according to this theory, as if they were immersed, involved and surrounded by a Higgs Field that offers resistance to the movement of matter, just like the same way as an object that moves in a fluid and feel forces of viscosity.

This particle has not yet been detected. And it is this particle that physicist want to discovered at the LHC, Large Hadron Colisor, in Geneva.

It seems important...

Physicists do work very hard and accelerate their research in order to find evidence of the Higgs Boson among the many traces of mixtures of particles collisions.

Higgs Boson has also been baptized with another name, the Particle of God. The reason why this happens is because it is general believed that this particle is closely linked to the origin of mass, to the fundamental constitution of particles, to the inner structure of matter... the Quantization of Matter!

This is a subject of great interest to all of us scientists and to any physicist, a subject which is still extremely difficult to explain: therefore, the limits of the final division of matter, the quantization of mass and its physical origin.

Let us first start by making a slight tour inside a Particle Accelerator.

Replications of Nature show us the same particles but with slightly different relative mass, thus, whenever an accelerator increases the energy it is obtained a Standard Particle Model, which corresponds exactly to the same particles with the same characteristics of electrical charge and spin, but with only one difference that is, heavier mass with more weight.

Whenever the energy increases ... it also increases the mass of particles. This situation probably happens, because energy must provide mass! If we are only increasing energy...

For this order, if we continue to increase energy, we still get the same particles, and possibly many other new and different particles, but still with heavier weights, more unstable and with a life time very ephemeral. Because the energy of our Universe is not suited to the existence of these particles...

In my opinion the Particles Accelerators create virtual Universes ... and contrary to what they wish to offer, these high energies are far to recreate the conditions of our Primordial Universe.

These repetitions of particles can be processed continuously and almost endlessly, but it requires some caution ... it is possible that a particle accelerator powerful enough, instead of creating a very heavy particle, can create a Black Hole just over the surface of the Earth!

It should be noted that there is an important idea to remember, all of these particles experimentally possible, are not the particles that exist around us, they don't share of our space-time. So, if there aren't most of these particles in the form of stable matter, if they don't have a free existence in Nature, personally, I do not understand how come it is that this extensive classification of many particles can bring great benefit to the understanding of our Universe...

This long trail of highly unstable particles detected in Particles Accelerators as a result of frontal shock between protons and anti-protons, electrons and anti-electrons, can release much energy and with it many new artificial particles.

For the moment it has been classified more than 400 of these new particles! This is a huge number of particles...

But the three fundamental and real quantum particles that form nuclei and atoms and the entire Universe, all entity with stable mass, since our own body to all objects that surrounds us, are just three: Quark up (top), Quark down (low) and the electron. These are the only stable particles with mass in this Universe.

According to Einstein's equation $E = m.c^2$ there is a proportion and an intimate relationship between Mass and Energy.

The first objective verification of this concept consisted in the following experience:

Starting by focus and fire a beam of accelerated protons on a Lithium sample, which is a light metal with atomic mass number 7, thus, in its nucleus are contained three protons and four neutrons. When the nucleus of the atom of Lithium was bombarded by a proton, this Lithium nucleus was divided and split up into two new nuclei of Helium, consisting now of two protons and two neutrons each. That is, the total sum of the initial particles was equal to the number of final particles:

The Lithium nucleus (3 protons + 4 neutrons) + 1 Proton bombarded = two Helium nuclei (4 protons + 4 neutrons). It is understood that the bombarded proton then collided with the nucleus of Lithium, initially joined inside the nucleus and in a final stage produces the split of two distinct groups designated as Helium nucleus, or also known as alpha particles. So that the total number of particles remained the same, as it was expected.

However, there were some intriguing data in this experiment. The total sum of the masses of the two nuclei of Helium was less than the sum of the masses of the proton and the nucleus of Lithium measured initially. What this means is that in practice we have some missing mass!

Moreover, the total energy resulting of this collision has shown to be greater than the individual sum of the energies of the proton and the core of the bombarded Lithium! The advantage was then a gain of energy!

All experiments of this kind led to the same result and what we can understand from here is that all these collisions are the evidence of a fundamental transformation: cause a loss of mass and an appearance gain of energy.

This is the basic principle that runs a nuclear power building. For the process of nuclear fission of Uranium elements, there is a disappearance of a tinny amount of mass which is accompanied by a release of an enormous amount of energy!

The possibility of this transformation was theoretically expected and predicted twenty years ago, in 1905, by Albert Einstein, establishing this relationship in his famous expression $E = m.c^2$.

This equation is, nowadays, a final acquisition, however, we cannot forget that the reverse process may also occur. With this, I wish to say that Energy is also capable to turn into mass, the inverse transformation is also possible, according to the expression:

$$m = E / c^2$$

And this is the process that has been used in Particles Accelerators. Because the resulting mass only depends of the speed of the accelerated particles and of the energy of the bombarded particles.

It is very logical that every time when the energy increases it is also seen and detected a rise in mass!

What are the exact mechanisms by which this takes place, I do not know, I couldn't say ... this is the greatest secret of Nature!

As well as understanding and decoding in which way the electromagnetic process is made and what are the exact mechanisms governing the production of electrical charge, I also cannot say ... the elementary charge of the electron is another of the greatest secrets of Nature. Despite electrical charges determine the existence of an electromagnetic field, however, we do not know why the electric charge is present, what is its genesis, what is the exact process that can bring to life this permanent charges , we do not know the laws that govern them, their intimate behavior which allows the existence of this continuous electrical currents.

However, the fact is that matter concentrates a huge amount of energy. And the lower the size of this mysterious substance that we wish to observe, the greater is the amount of energy needed to be spend and therefore the greater the complexity involved.

This means that if we are to remove an electron from an atom, we must apply a certain amount of energy, corresponding to the energy of ionization. But if we are to remove a Quark from the nucleus of the atom, this process it is not so easy ...

In fact, what experience have shown nowadays is that we haven't been able to accomplish the nucleus ionization, it seems that we are still not capable to apply enough energy to ionize the core, that is, to remove a quark from the nucleus of the atom ... or maybe we haven't been applying the correct energy ... so that the existence of a free quark as never been seen. This shows that the power of connection between quarks must be very high!

The energy required to release electrical charges, electrons, from its atoms is about a few electron-volts. However, exciting a quark requires energy in the order of mega-electron volts.

The forces between quarks are much more powerful than the electromagnetic and therefore offer greater resistance to the excitement.

The fundamental limits of matter, its most minimal and small particles, designated as quarks, concentrate enormous amount of energy, a really huge quantity of energy...

In an analogy, with a very brief and very simple exponential sequence, we can remember of the value for Atomic Energy (Atomic Bomb), after that we could consider the Nuclear Energy (Hydrogen Bomb) and going on the following scale we might get an idea of how high is the Quarkonic Energy!

The boundaries of matter involve a large amount of energy, because limits of matter are nothing more than a concentrated form of energy. Inside this frontier, the matter consistence, is the purest form of energy!

Leaning at the table in which I'm writing on, the soundness of this table relies in an appearance, as a result of forces that bind the atoms together. It is not a result of the concentration of matter. This force of resistance is a result of an equilibrium and balance between atomic forces, between molecules that are strong enough to produce the consistency of all that surrounds us. But this table is composed, mainly, of empty space!

Our entire body is in fact composed mainly of empty space, which somehow it seems solid, concrete, material ... but in reality it is not.

You can't even imagine how empty matter can be ... this mysterious substance and its apparent density comes only from the field and energy concentration.

If we were to consider the actual amount of matter that fills a single atom, we will conclude that this 'object' is made of, essentially, over 99% of empty space!

And its constituent particles that seem apparently to be solid, are only a result of the scale in which we are trying to observe them. If we could observe them more closely, we will see that everything vanishes in a filled of forces that are unimaginable strong and intense ... because atoms are not 'things'!

But we are still carrying the preconceived opinion that matter is solid, and as such it must have a minimal measure, a quantum portion. If the charge of the electron follows a minimum value which is 'e', it is believed that, in a similar way, mass must follow the same minimum standard which is 'm'.

Therefore, we unceasingly search for the minimum and indivisible amount of matter. It is developed many efforts in order to quantify the old Gravity, the Quantization of Mass, the final limits of matter, because it is believed that everything in the Universe has got a quantum unit...

The quantization of matter, the basic unit of mass, they will not find it, because it does not exist … not in that way...

What evidence do we have to find that the material things are solid, made of a consistent substance and, consecutively, with a quantum unit?

If we really see with our mind how do we pick up an object with our hand ... if we really reflect on how this simple action takes place, we will discover the correct answer about the quantization of matter...

If you reflect a little on this subject, you will see that the answer to this question is only one: which is that, in Nature nothing can be touched!

Each particle is composed of an individual field, independent and very strong. We never actually touch the objects. We never really make a real contact with them at all.

Everything interacts through forces and fields. The touch of matter is an illusion, a forbidden possibility in our Universe. As such, the existence of dense and compact units of matter, the quantification of mass   ... would violate this major principle of Nature, because a field is composed by concentrated lines of energy, but these lines never touch, never cross ... and matter is the purest form of energy.

The best idea that we can have to imagine what is mass, is visualizing it as energy in motion.

If we relate Newton's second Law of Motion F = m.a with the formula of Electric Field F = Q.E, we can say that:

$$m.a = Q.E$$

$$m = Q.E./a$$

This will be more legitimate formula to explain the concept of matter.

We can say that the fundamental substance of matter is the state acquired by a charge submitted and exposed to an accelerated energy field.

This would be the best approach for the definition of mass. Therefore, density does not represent the amount of mass per unit volume. The density of matter represents the amount of energy that exists in a given volume.

The mass is a fictitious structure, a shield, a resistance, a surface field, consisting of energy in a dynamic equilibrium.

If by any chance we could interfere and disrupt this balance, the object would vanish in their place would appear a huge explosion of energy!

The quantization of the matter is not possible because there are no real units of mass, thus, there are no minimum units of matter! The material field itself is an illusion of Nature! Maybe this is an idea a little difficult to explain, because we are used to see things, the material objects ... in fact, it is hard to explain that they are not really there, physically and in a material way.

We can consider matter as dense energy holograms, as knots in the field.

According to Einstein's equation $E = m.c^2$, the theory tells us that energy has mass and that mass is a form of energy. This is so true, a grandfather clock that oscillates is slightly heavier than one that is stopped, because the kinetic energy of the pendulum has got mass!

Indeed, things seem to have mass because they have Energy in Motion. Objects do not have mass inside. Objects do have moving energy. As faster as something moves, the more energy it gains; and the more energy an object has, more massive it becomes.

This idea, that matter is a form of energy or a field of forces, is not new and was already present in many minds in the past.

In 1844, Faraday has demonstrated that his ideas were too much advanced for that time. His work was publically presented for the first time to the Royal Institution, consisting of two studies with new ideas about lines of forces and electromagnetic fields. In its vision of success he anticipated the prospect of Quantum Field Theory of the XXI century, considering a replacement for the concept of atom, arguing that there could not be a real difference between space and the atoms in space, defending that these are two versions and manifestations of the same substance and that both should be considered as mere concentrations of lines of strength, as knots in the electromagnetic field.

With his visionary theory, Faraday rejected the concept of ether as well as real material particles. And in its place, left us a picture of a Universe consisting of nothing less than a web of fields in interaction!

Another mind, also advanced for the time, carried the same idea as we can see by the following quote:

"The Theory of Relativity teaches that matter represents an enormous reservoir of energy and that energy also means matter. Under these conditions, we cannot qualitatively separate matter and field, because the distinction between mass and energy is not really qualitative (...) the matter exists where there is a large concentration of energy (...) the distinction between matter and field is quantitative rather of qualitative. "- Albert Einstein-.

It's often read that Einstein wasted the second half of his life, in which he did not produce anything new or interesting.

Once revealed such an important theory as the Theory of Relativity, nothing that could come after could be of a lower dimension or with a smaller size of importance.

We can imagine the great pressure that he was being subject while he was trying to finish on time the final details of his theory. Maybe we had not understood the message he wanted to send us but wasn't able to complete ... a message that he already knew its contents and yet didn't know how to show us...

"We can look at matter as the regions of space where there is an extremely strong field (...) the field is the only reality." - Albert Einstein-1938.

The basic idea that we all know and learn very easily, that in the common matter the origin and source of Gravity comes from the quality of mass, must be abolished and replaced by the New Theory of Gravity.

For all the reasons demonstrated here, I feel in the obligation to note that Gravity is not a quantum phenomenon, but an atomic phenomenon. Therefore, we must conclude that ... there is no need to build a Quantum Theory of Gravity!

This reasoning follows that if the Force of Gravity does not exist as original force of the Cosmos, there is also no messenger and no particle that mediates this field ... there exists no such particle as the Graviton!

As it is also impossible to quantify matter, because matter is energy...

And there are 3000 researchers in the Particles Accelerator of Geneva searching for ghost particles!

Chapter XVII

Wave-particle duality

1. STANDARD MODEL

"Imagination is more important than knowledge."
- Albert Einstein -

The Theory of String is a wonderful theory, absolutely wonderful! Considered by all scientists as the greatest theory of the century and an excellent example of human intellectual development...

...It is only sad that it just has links to Mathematics; Geometry; Differential Calculus and Algebra!

The current leaders of this theory describe the String Model as the greatest scientific theory of the XXI century that uses Mathematics of the XXII century!

This new Theory of the Millennium aspires to solve the following problem:

"Show the existence and mass range of the Yang-Mills quantum theory in Re^4 with a Gauge group, a Lie G compact group, non-abelian and simple."

This is, currently, the major Problem of the Millennium that absorbs hundreds of thousands of physicists and billions of neurons.

The resolution of this problem would have fundamental implications in the field of Physics. But first, it is necessary to find a general solution to the Yang-Mills equation.

If the Schrödinger equation is practically impossible to solve for elements with an atomic number higher than three, for example, for an atom of Lithium; the truth is that until today no one could find the solutions of these equations, since the degree of complexity increases considerably. Finding the solutions of the Yang-Mills equations would be the greatest achievement of the millennium in the area of mathematics, but everyone considers it a challenge too difficult for now.

Despite all the effort and all the skills developed here in the construction of this theory, if by any chance the solutions of these equations were found, what meaning would that have for the physicists?

It seems that, at least, all the mathematics conjecture that has been developed is not intelligible for us.

If you reflect about this for a moment, it seems incredible ... the greatest theory of the century is based on equations that no one can solve! Furthermore, the versions of the String Theory are so many and so complex that no one quite knows for sure what are these equations!

The Theory of Strings began as an alternative to classical physics by replacing the concept of elementary particles by one dimensional objects or 'strings' and establishes a set of rules to approximate calculate what happens with quantum strings when they interact with each other in different ways of vibration. And the mathematics that, initially, was a way to understand this Universe, recently became the only possible way to 'see' this physical world! But in practice there is no way to relate all of that math to any real physical processes ... so, please do excuse me for my naive question: what for is it being develop so much Math?!

But the physicists have not lost their faith and, therefore, they still continue to build mathematical conjecture upon mathematical conjectures ... without worrying too much about with its practical sense, with its connection to reality, with its physical meaning or with any experimental verification. They expect to have all their efforts one day clearly revealed and, only then, it will be completely understood the whole development of this rigours thought.

In my view, I would say that physicists no longer do Physics ... this science is confined to a whole mental process.

No need to be a dogmatic follower of Karl Popper's model, to remember the definition of an open science that considers experimentation and refutation. We should understand that what is at stake here is the concept of science! Science ... we should never forget this word.

Reviewing that, among more than ten thousand scientific articles presented and published about the Theory of Strings, so far it was not made a single prediction of this model that can be tested!

Since 1973 that practically there has been none objective development in theory of particles or quantum theory that has worth to earn a Nobel Prize!

The Strings Theory is immune to both experimental evidence and to theoretical refutation. This means that there is no possibility of verification of its accuracy or veracity either of its falsity.

I almost dare to say that it is a modern scientific theology, a beautiful abstraction of mathematics and metaphysics, very rational of course ... but still ... Theology!

I must confess ... I share not much admiration for the Strings Theory.

The more I look at nowadays Physics, the more I see how much they do understand about Mathematics. In fact, physicist do know a lot about math ... much of the work of Strings Theory uses very sophisticated mathematics. Specific fields of Math that even the most experienced mathematicians are not familiar!

Usually, the progress of mathematicians is advance and ahead of the physicist. Today, however, the mathematician are far behind, trying to be updated and assimilating new information developed by the physicist discoveries in new fields of pure math! ... A ridiculous situation, at least!

The Physics itself exceeded the Mathematics, exploring new concepts of pure math.

The promotion of this theory led us to a new way of doing science, the Strings Theory led us to a new path ... a path that seems to have no end. Because, as we continue to walk through this path, we cannot know for sure whether if we are on the right way or if we are on the wrong track.

What's more unusual is that this theory does not produce results capable of confirmation. Its major limitation is the lack of experimental verification, its inability to make predictions in the high energy physics and also at low energies physics of our daily lives.

This crisis in the greatest XXI century revolutionary theory is already beginning to cause some impatience and controversy among physicists. We entered in an impasse, caused by a theory that promises much but gives very little.

Finally, all the problems of modern Physics can be defined as follows:

There are problems of 1st order and problems of 2nd order. The first order, to be tested require huge energy, impossible to recreate because we do not have the necessary technology to accomplish such experience, thus, it becomes impossible to be confirm. The second-order involves theoretical concepts of mathematics so complex, with many parameters and variables, which results, in practice, in an impossible problem to deal and to solve.

Unsuccessful theories do not produce more information than what is introduced. This is the failure that affects the Strings Theory!

Therefore, I would like to make an appeal ... the physicists have lost their concept of what is be a physicist, of what is to do Physics. It is not the formulas and equations that lead us the way ... it is us that leads the way to a formula!

If we concentrate to much only in one single musical note, we can lose all the beauty of the symphony ... the specialists have acquired the ability to know more and more about one thing, until they know everything about nothing.

Between the note and the symphony, between the specific concept of nothing and the general idea of everything, it stands the wisdom of knowledge.

If Physics continues to evolve this way, it will be lost forever in a direction of a colossal abysm of mathematical conjectures and esoteric concepts with no physical meaning, far beyond the limits of science.

This critic to the contemporary Physics recalls that the basic understanding of our Physics, until today, it did not needed such a sophisticated Mathematic in order to be understood. After a while and sometime of dedication and reflection, the laws of Nature are discovered and developed without resorting to large numerical artifacts. And its comprehension, which initially appears complex and difficult, it shows easily to assimilate. It is understood, at the end, that the basic and fundamental concepts are quite simple and easily to comprehend, almost as if they were very intuitive and even obvious!

As I share immense admiration for this science of the Universe, for Physics, I would not like to see this discipline vanish and lost, so I come here to remember that there is a ritual in Physics which has been maintained for centuries:

The Physics, works with physics! This is an experimental science, quite different from mathematics. This last discipline only requires rigorous demonstration and logical argumentation, very clearly and accurate calculations. It is immune to experimental proof.

The new Physics Theory of Strings is continually challenging the limits of human understanding, developing models that we are not able to understand. Personally, I do not think that we have reached the limits of our capacity to understand the Universe.

There is a very simple explanation for all this history of Strings Theory, a scientific explanation and rather conventional … We must assume that the Strings Theory seems to fail as a major candidate for a Unified Theory, therefore, we have to work a little more, leave this old theory behind, now with 30 years of age, and start looking for a completely new theory.

The periods of crisis can be revolutionary and visionary par excellence!

Currently there are two types of researchers: some for normal science, and others for visionary science.

In the physical sciences, the pure thought may not be sufficient to reveal all the mysteries of Nature. Moreover, the experimental verification may not be possible, disabling the possibility of introducing more information.

If all these tools fail … there is only one still left … use the imagination to explain phenomenon that we do not understand!

Most researchers are suitable for normal science. Despite they have been the best students in their degrees and a PhD thesis; having been able to solve the math problems faster and better than anyone ... there is only one parameter that they cannot dominate so fast ... the imagination!

How a known scientist once said: "Imagination is more important than knowledge." – Einstein -.

The scientific creativity is a parameter that the manuals do not teach. But we are talking about scientific creativity and not about too much abstract and almost esoteric concepts. We must understand where the limit is.

I would say that the most known researchers process data, while others, very few, can think and make relations.

The advance of science is often made by the contribution of many physicists, working together according to the same theory, but it is also possible that a huge advance in science is given only with the contribution of one. Those are the visionaries!

The exploration for a new vision is based on very simple principles:

First, there is always an advantage to understand what seems not to be working. This is a breakthrough and an advantage. It allows so step aside a number of hypothesis and assumptions and adopt a different method of working.

Then, if the facts do exist, it means that are possible, so there must be a theory that is suited to the experience. And this is another step forward, because, at least we know that our problem has got a solution, unless, in the last case, there is no solution for our equation because the facts are wrong.

And the last principle is not to underestimate the elegance coherence of a logical Nature.

The elegance of a scientific theory, absolutely innovative and creative, is always driven by the concept of esthetics, beauty, symmetry and, above all, logic and simplicity. This may seem a little irrational argument, the more complex calculations appear to be, the more they seem to have greater power than those that are simple. But it is not so. The truth is, and to rephrase Roger Penrose "It is mysterious, in fact, how come something simple and with good looks is more likely to be true than what is ugly.".

The nature of a Unified Theory should be simple, coherent, logical and relatively easy to assimilate.

The failure of the Strings Theory is based on its own basic principle that it's established and in which it is based, which is:

All minimum entities of space and time are subject to quantum fluctuations, even the gravitational field itself. This means that the space field, apparently empty, is not really empty at all, it is never empty. The quantum space-time-energy is constantly boiling with ephemeral particles that appear and disappear in a shaking frenzy, surviving only a few little moments of time. The microscopic space is not a constant and static space, instead it is an interactive space with floating particles and waves of energy coming up and down, appearing and disappearing. But on average, both energy and mass take the value zero. And for the gravitational field it happens a similar process. Although the classic reasoning tell us that in an empty space, which means with the absence of masses, has got a zero gravitational field, in the quantum reality that is not so. Quantum mechanics states that this value is zero on average but that the gravitational field also fluctuates between waves of uncertainty values.

The Strings Theory incorporates the Uncertainty Principle of quantum mechanics and this one is not compatible with General Relativity in a microscopic scale, or in a quantum scale, or also called as the Planck scale.

Over the years this path proved to be full of dangers and this Uncertainty has become an obstacle that no one was able to overcome. However, there were physicists who accepted the Theory of Strings and continued with their daily work of research, that at the typical scales are far beyond from the Planck units, taking only a simple note that these two fundamental pillars of Physics, Quantum Mechanics and General Relativity, are at the bottom absolutely incompatible!

Others, however, did not resigned so quickly. They felt deeply unhappy with this conflict, arguing and pointing to an essential and critical failure of our understanding of the Universe.

Very well!

If the Universe is understood in its deeper and elementary level then it should be possible to describe it with a logical and consistent theory, that incorporates both quantum and relativity models uniting them in harmony, and not by an incompatible theory.

The physicists focused their efforts in order to unify the Restricted Relativity with Quantum concepts and with the Electromagnetic Force and its interaction with matter.

Through some early developments and strong inspiration physicists have created a new theory, nowadays known as the Restricted Quantum Field Theory, or also called as Quantum Electrodynamics, shortly known as QED, or simply like Quantum Field Theory, or even better known as GUT, the Grand Unified Theory!

It incorporates quantum characteristics, because it assumes the uncertainty probability properties of quantum mechanics. It is also relativistic, because it absorbs the restricted theory of relativity from the beginning, and, finally, is a field theory because it implements the concepts of quantum field and classical fields, in this case, the electromagnetic force fields of Maxwell.

It seems that it is an excellent way to begin a new theory that aspires to contain all natural phenomenons!

Magnificent!

The Quantum Field Theory shows up as an alternative to Strings Theory and the truth is that it has achieve great progress in the direction of the unification of different types of particles with its mediators of their field interaction, making a correspondence to the different forces involved. It seems that this theory has composed a Standard Model that organizes three families of particles and three non-gravitational forces: the Strong Force, the Electromagnetic Force and the Weak Force. It has also been able to accomplish a more intimate interaction between the Weak Force and Electromagnetic Force, uniting these two forces together into a single force: the Electro-Weak-Force.

The Strong, Electromagnetic and Weak forces allow a proper interaction and accurate description in quantum terms. In fact, they are formulated in terms of a mathematical theory of quantum fields experimentally tested with a high level of precision never achieved before. Which is an excellent step towards a Unified Theory.

The greatest difficulty of this model relies, again, on the incorporation of Gravity. However, in our case, the problem that has been presented in relation to this global unification of the four forces of Nature and to its main obstacle, which is the fact that it has not yet been formulated a Quantum Theory of Gravity, is no longer a problem!

... Very interesting indeed!

By this moment, the new Model of Modern Physics fits only Three Forces of Nature. It includes the individual mediators of the interactions, and the particles of matter, which can be summarized as follows:

STANDARD MODEL

MEDIATORS OF THE INTERACTIONS

GLUON	PHOTON	BOSON	BOSON	HIGGS
g	γ	$W^{\pm}$	Z^0	Φ

PARTICLES OF MATTER

LEPTONS

e electron

μ muon

τ tau

ν_e electronic neutrino

ν_μ muonic neutrino

ν_τ tauonic neutrino

QUARKS

u up

c charm

t top

d down

s strange

b bottom

- Current Standard Model -

- Mediators of the Interactions and Particles of Matter-

However, Quantum Mechanics had already developed new perspectives and new concepts quite innovative. Considering, for example, that matter is composed of elementary particles but that they also have properties similar to electromagnetic waves. So that, particles and waves share common characteristics. In fact, everything that appears in our Universe can be summarized in two concepts: Light and Matter. And this interpretation of the principle of wave-particle complementarities allowed opening the horizon. Therefore it is immediately inferred that, if we wish and ambition to achieve a complete unified theory of light and matter, this it is will only be possible if we look at the particles of matter and at the particles of radiation as something similar, therefore, satisfying the same properties of transformation.

The interpretation of wave or the interpretation of particle must share a common origin. In other words, particles of matter and particles of radiation should be described at the same way: in terms of fields.

We can also imagine this new field has having both characteristics, but only different manifestations and possibilities of expression.

Only this way we can consider electrons and photons and other particles on an equal stage, like particles which obey to the same set of the Field Equations.

In the area of the Quantum Field Theory, particles are also considered like some kind of field, and common particles of the classical physics are defined as local densities of energy, where there is a concentration of a field of energy. An idea that is not new, as we have seen before.

With this form of thought it is intended to embrace both corpuscular and wave concept of particles of matter and particles of radiation. These concepts were highlighted in the theory of photoelectric effect proposed by Einstein, and in the wave concept of Young experience.

The most notable example of this theory, that corpuscular particles have got a behavior similar of radiation, and vice versa, consists in a very interesting experience of wave-particle duality. An experience very difficult to understand at a deep and intuitive level, this feature so surprising and amazing of the microscopic world!

Chapter XVIII

Wave-particle duality

2. EXPERIENCE OF DOUBLE-FISSURE

"The pessimist complains about the wind; the optimist expects it to change; and the realist adjusts the sails. "
- William Ward -

Let us imagine an electromagnetic wave, a beam of light, focusing on a screen where there are two fissures. When the wave passes through these holes, each one of them will be a new source of a wave motion, such as a material wave, like water, for example.

A key feature of this undulatory movement is the phenomenon of interference, which reflects the fact that the oscillations from each source of material wave can be added or subtracted from one another.

If we put a second screen, after the first one, so that we can detect the intensity of the electromagnetic wave that hits it, we will see an image of interference, resulting of a composition of alternating maximum and minimum, corresponding to a pattern of lighter and darker lines which are called as interference fringes. This phenomenon matches with the wave interpretation of Young's experiment and has proved the wavy nature of light.

If you repeat the same experience with material particulate, shooting bullets, for example, we can very intuitively deduce what will be the pattern formed on the last screen. There will be bullets that pass through one fissure and bullets that pass through the other fissure, so that the ending result is the concentration of particles/ bullets in two specific and distinct directions. In this case, there is no phenomenon of interference. This is, therefore, the result expected by classical physics, corresponding to Newton's corpuscular interpretation.

So far, so good, but only until now...

Because if we try to do this experience with other kind of particles, like electrons, for example, being thrown against the screen, with both slits open, it will be formed at the last screen an interference image!

Supposedly, we would think that the electrons will pass through one or other fissure, and it will be expected for them to form on the last screen a standard corpuscular concentration of particles, like the same case of the shooting bullets experiment, but, in fact, that is not what happens. Surprisingly, what is seen is the same phenomenon of interference, which is a property of waves and as such we must assume that the electrons also have wave characteristics. But how is this possible?!

This experience becomes even stranger when we try to conduct the same experiment but only with one particle, starting by fire up a single electron against the screen. First, we can see that there is a single particle that hits and reaches the last screen in only one point, in a very specific direction.

But when try to repeat this experiment several times in succession, releasing one electron at a time, one after another, each electron reaches the screen in a different point, however, when we overlap all the results together it is obtained, amazingly, the same figure of previous interference!!

Magnificent!

How come individual electrons passing through a slit or other randomly, can conspire together to form an image of interference?

This is one experience that can turn our neurons in to ashes ... how is this possible?

Is there a logical way for this to happen and be so?!

Now, let us perform the same experience with photons, imagining that we can reduce the intensity of the beam of light until we can throw individual photons against the screen, with both slits open.

Initially it appears that each photon reaches the screen in just a final point, and always in different points, in a very specific localization, just like if they were corpuscular particles. So we can deducted that the photon passed through one slit or the other, but it did not pass through by both fissures simultaneous and at the same time, which shows the corpuscular characteristic of light. But if you wait long enough, the photons will pass through one slit or another, but at the end of many passages it will also be form a picture of interference!

But if we close one of the holes and begin to shoot the photons, one by one, we will get back the standard corpuscular and located concentration of particles, just like if we were shooting bullets with one hole open.

The fundamental issue of these experiences is that they allow us to characterize both particle of radiation and particles of matter as having both corpuscular and wave characteristics and therefore we can establish this feature of wave-particle duality.

To get a clearer idea of this experiment, we will carefully view the following images which constitute evidence about this phenomenon:

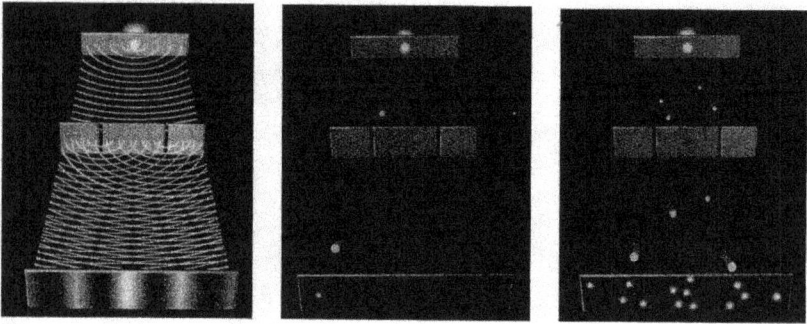

- Experience of Double Fissure -

- Behavior of any wave and corpuscular particle:

photons, electrons, etc. -

The wave particle duality can become even more exotic. If we try to determine from witch fissure did the electron or photon passed, we change the final result of the experience!! The figure of interference is destroyed leading to only a concentration of particles very well located!

This phenomenon, as a whole, is truly amazing, and gives us a lot of thinking! Is it possible to deduct that when we try to perform an experiment that shows the corpuscular nature of matter, that is, that locates from which fissure did the particle went and pass through, we completely destroy its wavy characteristic?!

158

The Quantum Mechanics states that we interfere with the alternative that the system chooses! Supposing that, in the microscopic world we have to ignore our intuition and all the classical physics concepts. So there is a well-known statement of Richard Feynman: "No one understands Quantum Mechanics.".

Compared with the clear and logical construction of Newton's laws of motion, or the electromagnetic theory of Maxwell, we can say that quantum theory is in a chaotic state, almost impossible to understand.

The Quantum Mechanics does not give us a very efficient description of these experiences. Despite the success of this theory in other fields, for example, the Schrödinger equation, which describes the physical states of a particle and its temporal evolution, later updated with the Dirac equation. It is this equation that replaces the Newton's equations in classical mechanics for a mechanics applied to particles such as photons and electrons, accurately defining all their characteristics and properties. However, for this case in particular, Quantum Mechanics does not feel very comfortable with the strangeness of their own experiences.

But is Quantum Mechanics so different from Classical Physics?

Many will say, without a doubt, that the answer to this question is a round and absolutely yes.

Allow me to say that I completely disagree!

The Quantum Mechanics presents us with very different concepts. Between probabilities, uncertainties, wave functions, and the wave-particle duality, all of that are quite radical new ideas that transform our vision of reality, capable to lead any classical and traditional physicist in to dizziness, hard to convince and to convert. These principles capture the essence of Quantum Mechanics. Properties that normally are above any suspicion, for example, the definition of the position, speed, time and energy of a quantum object, is now seen as mere fluctuations, uncertain, undefined probabilities.

The quantum physicist advise us to accept Nature just like the way it is. However, we must not fail to fall in to this current of mystical connotation and indetermination concepts of the quantum world, judging, from the beginning, these experiences as instantly illogical and impossible to understand. Some also give quantum science other descriptions, more dramatic, such as bizarre, absurd and irrational.

We should not let ourselves be seduced and surrender completely by these new concepts, without even trying to fit them into a coherent relationship.

The compromise between the concept of wave and corpuscle becomes a relationship almost impossible to assimilate and overcome, a concept incompatible with the human spirit.

My own conviction is based on a profound belief in the cause-effect relationship. I admit that I may be wrong, but personally I do not believe in the uncertainty and undefined concepts of Quantum Mechanics.

It seems to me that, and to quote the physicist Ernest Rutherford: "While we are not able to explain something in simple terms and not technical, it means that we did not completely understand the phenomenon.". I think that we still understand very little about Quantum Mechanics. As it still is necessary to explain physical processes in terms of technical and mathematical calculations rather than words and concepts, it can only show how far we are to properly assimilate the whole phenomenon.

The advantage of quantum mechanics is relying on statistical methods, without this tool, quantum mechanics is uncapable to explain a process or a phenomenon. The treatment of a phenomenon is seen when comparing to a large number of particles, and their statistical analysis allows a probabilistic forecast quite correct. A bit like the thermodynamic analysis of particles of gas in the atmosphere. It is impossible for a meteorologist predicts accurately the movement of all single particles in the air, but it is possible to obtain an analysis of the global movement of all particles, through a statistical and probabilistically treatment.

Thus, Quantum Mechanics works when applied to a large number of particles, but, unfortunately, it cannot be apply to treatment of a single particle. Because, for these individual cases, Nature has not yet revealed its greatest mystery!

The experience of the double-fissure exposes this mystery into the light, making it very clearly and absolutely intrigued the physicists. The fact is that no one can make an objectively analysis and enlightening explanation about this phenomenon, and this leaves all physicists absolutely and completely unarmed, without any chance of explanation, the only thing that it can be said is that we simply know this is how it happens.

The orthodox interpretation of Quantum Mechanics states that before the operation of any measure we cannot talk about realities because all that it exists is potentialities.

Before we continue let us see, in general terms, which solutions were presented for these phenomena.

Quantum Physics does not succumb to the tradition of Classical Physics saying that a particle passes through one slit or through another. It says that the particle passes through both!

Nor does it refer to the tradition physical concepts saying that a photon travels through one path or through another path. Quantum Theory says that the photon travels through all infinite paths simultaneously!

They wish to convince a layperson that one single particle passes through different paths simultaneously, and nothing less than an infinite number of them!?

And still that the relationship between the observer and the experience changes the final result of the experience. If the observer tries to intervene in order to understand and confirm from which slit the particle passed through, the system realizes it, it cheats, and chooses a slit defined!

I would personally ask for the Claims Book. This is absolutely insane!

The quantum interpretation rejects the concept of physical waves and only considers waves of probability. This means that, necessarily, to explain this experience we must abandon the concept of causality.

In my point of view if we reject this causal relationship, that involves giving up our ability to think about things and physical processes.

If we cannot establish a rationale thought based on the cause-effect relationship, in which way do we intend to make progress in science and consolidate new knowledge?

Personally, I think, both Strings Theory and the Uncertainty Principle have weakened the foundations of Science, in different and distinct ways.

If the Strings Theory questions the relationship between theory and experiment, the Physics itself as an experimental science; on the other hand, the Uncertainty Principle leads us to a destination without direction, because while it opens the doors to these new concepts of uncertainty, it closes the concept of certainty.

From here now it becomes almost impossible to make new science. If science loses consistency of theory with experimental verification, if science abdicates the cause-effect relationship, in which level is Science placed?

The major issue here is, again, the fundamental concept of Science!

The science of the old traditional scientists, worked with classical physical concepts, has always been a good guide, and has always given its fruits for those who were patient t and for those who truly believed...

Back to the duality:

Quantum mechanics claims and argues for the probability intervention, ensuring that the particle did not quite split into smaller pieces, it only indicates the existence of regions where the particle can be found with higher probability.

What in practice does not explain very much about the physical process, the fundamental part of the experience that we all would like to know ... the 'how'!?

Physicists are already very familiar with these experiences but, in my view, they are still waiting for an explanation.

Let us focus and concentrate ourselves only on the description of the facts. And it's worth to pay much attention on the details. If indeed this experience happens, then, it's because it is possible. We just have to know how!

I have a suggestion, not very accurate but it can raise another hypothesis ... more classical!

I would like to have the opportunity to emphasize, once again, that Modern Physics should not forget and exclude all the principles taught by the modest Classical Physics. This old science accumulates immense wisdom and has not yet revealed all its knowledge.

There is a law in the Universe in which I still cannot quit believing that is the cause-effect relationship, and this is the fundamental pillar of the whole construction of Classical Physics.

How someone would once say "God does not play dice." - Einstein - therefore it is obvious that Nature have found a way to manage the duality in a logical process.

If you take an overview and look at the complete picture, we can notice that we have three major problems in modern physics which are waiting for a resolution: The experience of wave-particle duality; Intervention in the double fissure; and the electron quantum jump. Looking closely we can see that all of these problems are waiting for a common explanation, because they deeply are all effects of the same phenomenon!

First, let us go to the enigmatic quantum leap of the electron:

In a first approach, apparently magical, we can say that, when photons collide with the electrical atmosphere of an atom they materialize the electron at that point, making it in a corpuscular particle. The action of the photon at the periphery of the atom establishes the concentration of the electrical charge density of the electron in to a specific area and location, according to its corresponding energy level.

The principal quantum number that sets the radius of the orbit where the electrons can remain have a specific definition with discontinuous ranges, because they define the distances in which the emission and absorption of energy can take place, thus means the energy levels of the atom. Only in those specific orbital and only in such moments is when the electron takes its corpuscular behavior. With the absorption of energy from the photon occurs the collapse of the wave of the electron. In this process the distribution of the electrical density is transformed into a concentrated point of electrical charge, thus, the electron becomes a corpuscular particle.

On the other side, we have the opposite process: with the emission of the energy from the electron occurs the collapse of its corpuscular behavior and the electron spreads apart into a cloud of electrical density, spreading into a wave of electrical charge, being no longer a concentrated corpuscular particle.

However, initially, the electron has got its density of electrical charge more or less evenly distributed surrounding the atom, according to the shape of the orbit around the nucleus. In the presence of the photon the distribution of this density is changed and it is then when it seems to occur the quantum leap, which in reality is not exactly a jump but rather a linear change of the distribution of the electron density charge, which becomes concentrated into a single point, with a radius distance from the nucleus very well defined, corresponding to the energy level. And then we say that the electron described a quantum leap, but in fact what happened was that the electron assumed its material appearance.

When this happens it is because the photon has got the power to act upon the charge of the electron, concentrating it into a single point, transforming it into a corpuscular particle, so that we can consider the photon as a particle carrying out its own field which will have action on the attraction of minimum amount of electrical charges, in this case, the negative charge of the electron.

We could assume an analogy, just like the Strong Force that attracts the positively protons charged in the nucleus, keeping them together, the photon would have the feature to attract, unite and concentrate low densities of minimum charges, joining and keeping together the electrical charge of a single electron, the elementary charge.

In this first stage, let us keep only the major idea of this concept, which is that particles in their natural state of balance always assume their wave status but if they are disturbed by a photon they will change their form and assume their corpuscular appearance. The intervention of the photon is crucial.

Now, let's see what is happening in the double fissure experiment...

The conditions of the experiment are very important. The experiment is performed as follows: isolating the system from outside light, that is, placing the emitter of particles, the bulkhead with two fissures and a final light-sensitive screen in a closed system, isolated from the intervention of external photons, since what we want to check is the pattern of light and contrast that is formed in the last screen, just like a photographers film.

When photons are project, the photographic plate records the light that passed through the slits showing clear points of light. When particles such as electrons use projected it is used a phosphorus screen, this substance is sensitive to any external action, and is basically the same principle that is present in old televisions with cathode ray tubes, and he shock caused by these particles will also cause bright points.

Having this clear, we can move on to the first phase of the experiment: the projection of an electron beam with two fissures open:

When the electrons are projected against the material target, against the screen, they gain speed and as such they assume their wave nature. Since during this journey there is no interference with photons, the electrons do not suffer any disruption and once they have not been disturb by any photon they continue their path in its wave aspect. The wave nature of the particle is then preserved.

As we have two slits open, the wave nature of electrons gives rise to the phenomenon of interference. The result in the final screen is the projection of waves.

Second phase of the experiment: the projection of a light beam with both slits open. This experience is virtually identical to the above:

The light is an electromagnetic wave, consisting of a huge number of photons, just like the waves of water are formed by a huge current of H_2O molecules. The wave nature of light is maintained from the beginning until the end and, as expected, it is formed the phenomenon of interference in the final screen. The result is the projection of waves.

Now, in an attempt to keep this line of thought, we will move forward into the part of the experience where we want to know in which fissure did the electron passed through.

Of course, if we wish to know which fissure is that the electron passed through, we must point some light on it in order to see it. When we project light on the electron this one reflects its presence and position, as any other object. But, this intervention by itself causes the necessary interference to change the aspect of the initial wave appearance of the electron. The contact of the electron with the photon transforms it automatically and necessarily in its corpuscular aspect, for the reason that has been explained in the quantum leap.

There is no direct relationship with the observer, there is a relation of cause and effect! The intervention of the photon is fundamental.

Now, the trick of the magician is at the end of the experiment, at the corpuscular property of the electron and the photon!

The light shell still contain many properties that we don't know and certainly many secrets to reveal.

However, in the double-fissure experiment we have been paying attention only to the first part of the experience, thus we have been only focus on the part of the action and we are forgetting of the 3rd law of Newton about the principle of action-reaction!

The final screen provides a reaction, reaction which modifies the final destination of the experiment. Let's see how:

We know that when a material wave is spread towards an obstacle, for example, we can consider a sound or water wave, we know that there will be a reaction, the wave can be reflected in whole or in part, because the barrier offers a reaction.

Could it be that in the case of the projection of one individual particle, whether it is a corpuscular or a radiation particle, our obstacle also offers a reaction?

And if it does, what kind of reaction could that be?

We must first clarify what we mean by 'particles' of radiation or by 'corpuscular particles'. I can already say that the corpuscular concept is the one which impairs our reasoning.

Let us move first to the experiment with only one electron.

For the specific case of one electron, I must confess that, I was initially thinking about the phosphorescence property of the material and on the reemission of light and, therefore, on the action of photons. But this would only make the problem even more complex and reduce the chances of achieving the exact same standard image in the final screen. Then it came out another option, much more simple!

After we have in our hands the solution of a process, it is only then that we realize how simple it is...

It is good to think that things and processes are simple, this is always a good guide. But sometimes they can even be annoyingly simple!

Again, the photons always have its primary role, they decide the final outcome of the experiment.

As should be recalled, the electrons do not only emit photons when they are transitating energy levels. There is another way to produce the emission of photons. And what is that other way?

I leave you to think for a moment...

168

The solution to this experience comes from the following relationship:

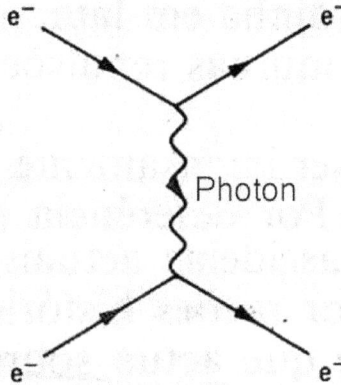

- Feynman diagrams! -

With this concept in mind, we can proceed with some more information:

When two electrons are on a route of collision they communicate with each other repulsive, informing and warning the other electron of their presence and proximity. Since the nature of these particles is naturally repulsive it starts up a 'fight' between them with the transmission of photons against each other.

In a way, the photons act as a messenger that carries out the information from one electron to another, communicating the following message: 'do not get near me, our charges are repulsive. "

With this concept in mind we can proceed with some more explanation.

Recall, first, that the initial movement of the electron is always a wave. Whenever the electron starts in motion or is moving with velocity it assumes its natural facet which is undulatory.

Only at the time when there is the collapse or spreading of the wave, which corresponds respectively to the moments of absorption and emission of photons, and only in those cases, the exact moment of collision or the collapse, is that the speed of electron is changed and, therefore, the particle assumes its dual nature permutating into a corpuscular or undulatory form.

In this experience of the projection of an electron at a time, with both fissures open, the final result, initially, stars by being corpuscular but at the end we will have an undulatory composition. Let's see how can we paint this picture.

First part:

What happens in the passage through the fissures of the first screen? The electron passes through them like a wave or as a particle?

We have seen before that quantum mechanics considers that the particle passes through both simultaneously, as if the electron had the gift of ubiquity.

One step at a time, and once again it is necessary to gain some degree of abstraction: First we must assume that the electron travels in its wave aspect, since that is that their natural state and, therefore, it approaches the first screen, consisting of two fissures ,in the form of a wave.

And what happens when the electron reaches the shield of the two fissures?

Although the electron initially travels toward the first screen in the form of a wave, it does not pass through both holes like a wave, that is, simultaneously. Once the wave front approaches the screen and confronts the shield, and whether if the electron is not immediately absorbed in the first screen, there still is a slightly chance that the front wave of the electron that is moving is slightly closer to one fissure or another.

Remember that when a wave is spread it has always got a circular design with a constant radius around a central point, therefore we can consider the advance of a front wave.

Once this front wave is sufficiently closer to one of the fissures it stars the recognition of the possible collision with the electrons of the plate closer to the fissures and, as such, the electronic wave of the electron collapses and it is transformed in its corpuscular aspect. It is on this stage that the traveler electron exchange photons with the electrons of the plate surrounding the fissures.

Summarizing, being the fissures of a small size, the wave front of which is closer to get through one of the slots should be close enough to recognize the electrons of the slot itself. As such, there is an exchange of photons between the electrons around the fissure and the traveling electron. The electron takes its corpuscular performance, and this fight avoids the route of collision between the electrons of the plate and the electron traveler, so that the electron can 'turn aside' and overcome the hole of the slot with safety, the dispersion of the electron avoids the frontal shock with the first screen.

As soon as it gets through the fissure, our traveler electron is now far away from the first plate and, thus, the traveler electron can quietly continue its journey assuming, again, a wave facet. With the particularity that the exchange of energy with photons has changed its momentum and direction. For that reason, when new front wave advance they always get spread in different directions, moving towards the final screen with distinct courses of collision, shocking the final screen always in completely different points. And, again, before colliding with the final plate, the fight with photons stars over again and undulatory and corpuscular permutation repeats.

Second Part:

Let us now focus only on the final plate.

When it is near to the final shield, the new front wave that comes, which carries another momentum and direction, begin to detect electrons nearby and the possible collision with the electrons of the final plate.

When both electrons are close enough to recognize each other start up the combat with photons. And where there are photons in the proximities, what happens? The collapse of the wave of the traveling electron and therefore it reaches the final screen in its corpuscular aspect! Since there are no fissures where our traveler electron can escape, the collision is inevitable.

In the case of emission of several electrons simultaneously, as I mentioned at the beginning, the end result is the image of interference, however, the situation is the same as the emission of a single electron. There are always waves on the way but they always crash on the screen in a corpuscular form. What is happening is that, given the speed of displacement of these particles, whose values are on the order and magnitude of the speed of light, around 10^7 m/s, the process of corpuscular overlap is so fast and it happens so quickly, that induces our eyes the feeling of having arrived a single wave simultaneously. But that is not quite what happens.

Let us see this process closer and in slow motion …

The first front wave moves with speed, the electrons of the final shield begin to feel the threat and start preparing their defense. The Soldiers of the Command of the front wave that is approaching make their first assessment of the impact territory and detect the enemy, the Elite's

Soldiers of the final shield. When the front wave is sufficiently close enough, it begins the attack without mercy with shots of photons. The electrons take their position and assume their corpuscular form, continuing the fight. Meanwhile, more Soldiers of the Command come on the way, new front waves advance and move quickly, taking the opportunity because they know that the Elite's Soldiers of the final shield are still busy with the first soldiers of the command. The Soldiers of the Command are in majority and the Elite's Soldiers of the final shield are powerless for so many invaders, that is when other fronts begin to arrive, to distribute, and take possession of the territory in other areas, away from the center, reaching the target in other fronts.

A plan of attack that not even Napoleon would remember!

And that is why the interference figure is slightly brighter in the center and softer on the edges!

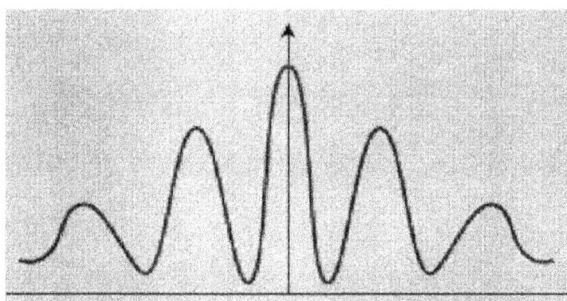

- Graphic of distribution of the spatial energy intensity on target. -

For the experiment with a single photon, the phenomenon is identical to that of a single electron.

The subtlety of the reasoning is the following: Everything starts as waves but everything ends as particles!

As you can see, the wave-particle dualism does not carry the uncertainty of position and velocity, what it does carry is a complete unknown of a process or a phenomenon.

When speaking about electrons, photons, protons, etc., we must say that these quantum entities propagate themselves as waves but exchange energy as particles! And we can consider that this exchange occurs in the form of packets of energy, quanta, or as discrete quantities of energy fields.

Chapter XIX

ANNIHILATION OF MATTER AND ANTI-MATTER

" While Men argue, Nature acts."
- Voltaire -

Another complicated issue that arises in cosmology is an old problem of Physics so fascinating and mysterious as the dual wave–particle duality.

The evocation is as follows: We now know that the Universe is composed almost entirely of matter, but it is concluded that it has not always been that way. In its remote past, the Universe contained almost as much matter as antimatter. The antimatter has got symmetric properties of matter; if by any chance two such particles meet each other they will be instantaneously annihilated. And as a result of their collision and annihilation we will assist the emergence of an enormous amount of energy.

The anti-matter was postulated by Paul Dirac, as a result of an equation which he formulated, the Dirac's equation:

$$[Y^\mu\,(\,i\,\frac{d}{dx^\mu} - eA_\mu\,) - m_e\,]\,\Psi\,(\,x^\nu) = 0$$

This equation intended to unify the Schrödinger equation with the restricted relativity, describing and joining both properties of particles: the quantum properties and also their relativistic properties. The Dirac's equation, in fact, solved the problem, or almost, but gave us something new.

Unexpectedly, from the solutions of this equation arise some new particles, different from the other known old particles. But curiously, these new particles were not so different after all, their properties seem to be exactly symmetrical. This means that, for each type of particle that we know in Nature, there is an anti-particle of equal mass but with opposite sign of charge. It follows that, for example, the electron has got an anti-particle called 'positron', of equal mass but positively loaded!

The existence of antimatter has been experimentally verified in particles accelerators. Our antiparticles really do exist, but it is good to know that they are not out there on the loose!

The theory of quantum field explores the symmetry of particles. It considers that in the beginning of the Universe, and considering the Dirac's equation, there should be the same number of particles as anti-particles. Deducing that antimatter has evolved together with matter. However, somehow, matter has prevailed over the anti-matter.

If matter and antimatter existed in an early stage of the universe, the situation would be truly catastrophic. All the matter and antimatter would interact each other and disappear, being mutual annihilated, leaving behind a trail of energy and radiation … and nothing about matter!

So why are there particles of matter instead of nothing?

Placing the question this way it becomes very hard to give an answer … it really seems very difficult to find a solution … very difficult in deed. No one has yet imagined a process that may have separated matter of antimatter; or a process that may have offset the increase of matter rather than antimatter.

Physicist talk about minor fluctuations, dissymmetry, minimum excess, lowest probabilities, processes involving different speeds of reaction of matter in respect of antimatter; or that the Universe still has got this antimatter hidden somewhere in the confines of space, so that the quantity of matter and antimatter is still evenly balanced until now, with equal quantities of mater and anti-matter! ... It really seems a highly complex problem.

Matter dominates antimatter but why is that so?

Researches exhaustively work, in order to find the solutions and explanations of why don't we currently detect the existence of all this antimatter!

This problem takes to the beginning of the Cosmos; to a specific period of the life-time of Universe called as the Hadrons Age or Hadronic Era.

It appears that our Universe, with only 10^{-7} seconds old, was composed of a set of heavy particles designated as hadrons. Given the extremely high temperatures deemed to be at that time, these hadrons would disintegrate into their fundamental constituent particles, the quarks. So that, the Universe would be filled by a primary plasma of quarks, gluons and high energy photons. The passages of this high energetic photons lead to the creation of antimatter.

Currently you can assist the creation of antimatter in Cosmic Rays, produced in nuclear reactions of stars, like our Sun for example. The cosmic rays represent high energy radiation, gamma radiation; and antimatter can be created by the passage of this radiation in space. The passage of this energy in the vacuum space does arise, spontaneously, a pair of matter-antimatter: the electron-positron pair. Happening that this matter and its homonymous antimatter have an ephemeral lifetime. The antimatter does not 'float' freely in space, it disintegrates in a fraction of a second.

Particles Accelerators can reproduce these events through frontal collisions with fundamental particles, which are quickly annihilated and turned into pure energy. Consecutively, this highly energetically radiation produces matter and antimatter

The theory of the Creation of the Universe, the Big Bang, believes that this process of creation and annihilation of particles actually occurred. However, it does not explain how come particles of matter managed to escape at this early period of annihilation. Considering that, in its remote past, very energetic and very hot, our Universe would not have a chance to escape to such a devastating annihilation!

Sometimes, when we cannot solve a problem we still do have a last resort ... we can try to change the equation!

If you still remember of the model that I have presented you about the prime time of evolution of the Universe, we have seen that the formation of quarks took some time to be develop; the gradual evolution of matter also needed some while to emerge; and certainly the appearance of photons associated with the formation of the electromagnetic force, happened in a much more later period. And the formation of the first atoms occurred in a period which the conditions of temperature, energy and radiation of the Universe were much lower, balanced and stable. In these conditions, with thermical balance and lower temperatures, the passage of photons and the high energy collisions between particles does not happen so often, nor have enough energy to produce antimatter.

At the beginning of the Universe we cannot consider the existence of cosmic rays; we cannot consider the existence of gamma radiation; not even electromagnetic radiation; much less the photons and convenient hadrons...

One of the major arguments for the existence of this annihilation process arises from the background radiation which has detected an excess number of photons in terms of baryons or common matter. And it is believed that the reason for this imbalance, between photons and matter, are traces of a process of matter and antimatter annihilation which occur in the past, leaving behind a deficit of matter and a sea of photons.

But even when it is considered that the ratio between the number of baryons and the number of photons in the primordial Universe is quite unbalanced and that the density of photons exceeded the density of neutrons and protons, we cannot assume that the explanation for this excess of photons had origin in the matter-antimatter annihilation process, because this excess is insufficient and minimum.

This excess of photons arises from a later period of the story of the Cosmos, and comes from the first form of production of light radiation in a Universe still quite young and absorbed by high energies. In this Universe, we must consider and remember that the intensity of the forces of Nature is variable and, particularly, the intensity of the electromagnetic radiation is much higher at that time, as such, the production of photons is indeed quite higher at this period.

This excess of photons, probably had its origin, not in a process of annihilation, but rather in a process of creation, produced by objects capable of delivering large amounts of light ... the Quasars!

Thus, what we have here is not exactly a problem, but rather a problem in the formulation of the problem, because:

The creation of electron-positron pairs never occurred!

The annihilation of matter and antimatter never took place!

The Hadrons Era never existed!

179

…There never was a Hadronic Age!

The Primordial radiation, associated with the moment of creation, is not, in any way, related to the electromagnetic radiation.

This early radiation is not, by any chance, connected with some kind of combination of all forces of Nature or with a Unifying Force.

This single radiation, early present in the life time of the Cosmos, is associated with another Exotic Force, the Apeiron... soon we will see what are the origins and characteristics of this Fifth Essence.

Chapter XX

PROBLEM OF THE HORIZON

"Calculate what man knows and there is no comparison
with what he doesn't know. "
- Chuang Tzu -

The Problem of the Horizon has a similar resolution. Everything relies in the fact that we have considered that the four main forces of Nature are innate to the Cosmos. As it has been demonstrated, the main forces did not all emerged with the Big Bang, these fundamental forces also had an origin and a natural evolution.

For this particular problem, what is at issue here is the combination of photons with the period of Inflation. During this inflationary period, in which the Universe underwent in a rapid expansion, the creation of the space itself happened so fast that even exceeds the speed of light!

This quickly expansion creates gaps and little pieces of disconnected horizons, making impossible any explanation for a uniform and coherent physical interaction. If the space expands faster than the light spreads, automatically emerges regions not exposed to the light, thus, the whole space should be filled by regions with different visual horizons, that is, regions that cannot see each other. Being the expansion of space faster than light spreads, it will infinitely expand the distance that separates those points in the space from the rays of light that pass through, and these rays of light will never be able to achieve that space.

Looking at the Universe today, and so that light could uniformly cover its entire space, the speed of light should be considerably higher, mainly or almost infinite, and it is not. When, in fact, we contemplate

181

the sky on a dark night we do not seen empty regions in the darker sky, areas with large gaps of light. The telescopes and image satellites show us a picture of a deep sky very homogeneous, isotropic and perfectly harmonious.

The first evidence for this problem is that there are no disconnected horizons... there aren't any ... at least in this Universe. That is our greatest evidence!

It is true that the radius of the Universe itself extends far beyond of what we see or measure. But the problem that arose for the predictable existence of these regions in space, with small radius, distinct visual horizons, and the fact that we still did not find them, and the enormous period of time that has been given in order to find them and to justify why the Universe presents us only one distinct horizon, all of the efforts and the many solutions and explanations that failed but which still continue, might be easily resolved if we assume that there were no photons in the period in which the Inflation occur, thus, there is no Problem of the Horizon to be resolved!

Really... if we just change one single little piece of this enormous puzzle which is Physics, everything starts to run well and naturally, without obstacles, without any problems ... no inconsistencies! The changing and the introduction of this simple criterion facilitates the entire History of the Cosmos.

It's a little bit like the theory of the Inflation. We have no direct evidence that this event really happened, we just know that with the inflation we can solve all problems!

With these new data, if I may... I would like to have the honor to make some changes: As you can see, the following table shows us the most important events which marked the History of our Universe...

TIME	TEMPE RATURE	ENERG Y	UNIVERSE RADIUS	IMPORTANT PHENOMENON
0	?	?	≈ 0 cm	Big Bang
10^{-43} s	10^{32} K	10^{19} GeV	10^{-50}cm	~~Gravity is strong. It's still needed a Quantum Theory of Gravity.~~
10^{-37} s	10^{29} K	10^{16} GeV	10^{-33}cm	~~Grate Unifying Force. Strong Force; Weak Force and Electromagnetic Force join together in one single force .~~
10^{-36} s	10^{29} K	10^{15} GeV	10^{-15}cm	Inflation
10^{-33} s	10^{27} K	10^{14} GeV	10^{-10}cm	~~Predominance of matter rather than antimatter. End of Hadronic Age.~~
100 s	10^{10} K	10^{-4} GeV	10^{5}cm	~~Nucleus synthesis. Formation of the first Hydrogen and Helium atoms.~~
10^{6} years	10^{3} K	10^{-1} GeV	10^{10}cm	Photons dissociates from matter. Origin of the Deep Field radiation.
10^{10} years	3 K	10^{-3} GeV	10^{20}cm	Today. Formation of Galaxies and Life.
10^{40} years	?	?	?	Wastage of matter. Proton's Disintegration.

- Main events of the Evolution of the Universe -

From my point of view I have to admit, I do not agree on almost nothing with this information! But after making some changing... oh yes! It is much better! Simple, refreshing and without complications!

Chapter XXI

DARK ENERGY

"I live in a very small house
but my windows open to a very big world."
- Confucius -

In the field of Astrophysics it occurs amazing compressions of matter!

For example, when the stars exhaust all of their fuel, they begin to contract, and under action of its own weight they begin to shrink ever more, becoming into a fraction of its original size. Some of these stars suddenly implodes and form extremely dense stars, confined into a very short radius. We can imagine that stars larger than the Sun may be compressed up to the radius of the city of Lisbon.

We say that the gravity intensity which exists in those collapsed stars is so high that even atoms themselves are force to implode, being almost crushed. And also the electrons are forced to merge inside the atom, being pushed into the nucleus. The atomic structure is broken as a consequence of the 'gravitational' compression of matter that involves it. In this new type of matter, with no electronic orbital, no electrons around the nucleus, it is given the name of Barionic Degenerate Matter, and for this to occur it is necessary to concentrate 100 million tonnes of matter into a single cm^3!

When the electrons are forced into the atomic nuclei they combine with protons and are transformed into neutrons. This new type of matter, consisting entirely of neutrons, offers sufficient resistive pressure capable to stop the 'gravitational' collapse of the star. The star stabilizes, and we say that it has become a neutron star.

A neutron star contains, probably, the most rigid material of the Universe. But even this is not entirely uncompressible, if the star has sufficient initial matter, this material can be further crushed, and the star collapses completely. The degenerated nuclei of matter suddenly implodes in a fraction of a second, leaving behind a Black Hole!

The Black Holes are simply mysterious! Where did it go, all that absorbed matter?

Until this moment, there is no satisfying theory that reveals the mystery surrounding the fate of the whole matter that imploded! Is it just gone?!

If I can still remember, in the Universe nothing is lost, everything is transformed!

It is believed that the center of a black hole concentrates an infinite density of matter, considering that gravity in that region is also infinite. When an infinite appears in a physical theory, it becomes completely haunted! The physicist do not feel very comfortable with infinite numbers ... therefore, they usually do not appreciate black holes; or they don't seem to like very much singularities or infinite states of space and time, because under this framework it's not valid the Laws of Physics that we have.

For a long time, physicists have been skeptical about such physical entities; were reluctant to believe that such extreme situations of material objects could possibly occur. But the black holes were detected and, thus, they were not merely the result of excessive imagination of some theoretical physicists.

It has been given much thought about what is at the bottom of this well and very shaped funnel... This type of vortex shape that appears immediately in our minds send us a wrong idea and a wrong image about what a black hole is. I personally don't appreciate very much the funnel shape … I prefer the geometry of the spheres ... the 'Bubbles of Oil'!

As I said earlier, not everything that we learn in school is true. However, not everything is false! Far from it!

One of the laws that we have learned in school, and most likely it will not be wrong, is the Law of Conservation of Energy.

Looking at a black hole, suddenly and just for having a general view of this physical entities ... if it is supposed that gravitational energy in that place should be infinite, since the entire mass has been concentrated in a radius zero ... and looking around, we know that the gravity around these points in space is very strong ... but if the gravitational attraction was actually infinite ... we would not be here!

Looking at others black holes spread throughout the Universe ... counting all of them, making the accounts, adding everything ... if the matter has disappeared completely and there is no infinite Gravity, then there is a violation of the Law of Conservation of Energy?! Is there a leak in this major principle of Physics? A 'hole' in the Conservation Law of Energy? Impossible! Impossible! … I refuse to believe so!

This confirmation would completely change and reshape the Law of Conservation of Energy. I do not think that this statement is incorrect. But, after all, where does it go all of that energy from a black hole? To the end of a deep well?!

Our Universe is not that strange. It will probably be, above all, logical!

If we assume that there is no leak, then, there is only one chance left for us to try to save this immaculate law. If something is missing, then something else needs necessarily to appear!

There is a new thing that is constantly appearing, expanding and evolving, entering in our Universe constantly, filling all the space around us, it has an ubiquitous feature and it is just in front of us ... that is the Dark Energy!

The existence of this new form of energy has been recently proposed, in 1998, when acquired data about the separation speed of the galaxies did not correspond with the predicted data. According to the theory of Big Bang, the constant expansion of our Universe should be governed by a particular value, however, it appears that the rate of expansion is much higher. The Dark Energy emerges as a justification to try to explain this rapid expansion of the Cosmos.

We know that the Universe is expanding and that the origin of this expansion came with the Big Bang, which has projected all space and matter, a bit like a classical explosion. However, the energy of this explosion is not falling, is not slowing down. Quite the contrary, the acceleration and speed of expansion of the Universe is expanding!

Typically, the energy of a conventional explosion is lost, attenuated, diluted, until it ceases completely ... and never increases! In the case of the Big Bang, we are assisting to an explosion which instead of dissipating energy is wining it!

To avoid this paradox, it was suggested the existence of a new force responsible for separating the Cosmos even faster and it was given the name of Dark Energy, which reminds a little the lost 'cosmological constant' of Einstein. From where does it come this 'cosmological constant' that stretches and distend the fabric of space-time imperceptibly? We cannot count on it as originating in the Big Bang, therefore, it must come from somewhere else. Somehow this energy seems to enter in our Universe, but not in a constant way, this energy seems to be emerging in a dominant way, because this dark energy has state to appear in increasing quantities. This energy is indeed immense. To get a clear idea, 70% of our Universe is composed of this Dark Energy!

One of the main characteristics attributed to that dark energy is that it is naturally repulsive, or else, it acts as if it possesses Antigravity!

Unlike the classical Gravity, this new type of gravity does not attract, rather the contrary, the nature of this force is naturally repulsive, or rather, it does not contain gravity on its own! Indeed, it is a very unusual feature of the Cosmos: Antigravity. Most of the ordinary people never heard of such a force, however, it exists and there is only one way to get something that does not produce its own Gravity, and what is that way?

According to what we have seen earlier, the evolution of the classical Gravity is a late production of matter, which can only arise when the atomic structure is stabilized, so that, we can say: without an atomic structure and without an atomic electrosphere there is no Gravitational Force, and we can add up, if we wish, that at a quantum level the elementary nature of matter is Antigravitational. In the specific case of the complete Degenerate Matter it is marked by the collapse of its gravitational structure; by the rupture of the electrosphere of the atom, once the electrons are forced to fall inside the nucleus and the atom collapses on itself. So, after a while, this new type of matter will behave as if it had a negative pressure which produces natural absence of Gravity, losing its compression, assuming a repulsive constant. We can say that, this new form of matter-energy density will behave in a different way, assuming a negative pressure and density. And on the contrary of the positive density of common baryonic matter that we see spread away throughout the Universe, this new type of energy, which has a different density, will try to find a new way to break through, another place through space which it can escape. Therefore, we can say that this new form of Negative Energy flows 'out' of our visual Universe.

However, this energy does not disappear, but rather, involves us and surrounds us in all directions. If you noticed, when we first started our Natural History of the Universe and the moment of Big Bang itself, there was a new form of energy already present, a primitive energy, the Pure Primordial Radiation, which has, in its basic features, the same descriptions and properties of the Dark Energy itself. My question is: are these two types of energy one and the same thing?

If the characteristics of Dark Energy are so similar to the original energy in the beginning of the Universe, then, we might consider that they are the same form of energy. We could consider this form of energy as an elementary cosmic fluid, the base that allows and supports the development and evolution of everything, and hence the transformation of this energy into barionic energy; in the common energy which supports the construction of atoms, planets, stars and life itself. This would be the raw material of the Cosmos, the Apeiron of Anaximader!

And much more than a simple raw-material, this form of energy incorporates the very support of space-time itself! So that we can consider it as the Fifth Supreme Force of the Universe!

First, let me just mention that we should not confuse Dark Matter with Dark Energy, because there is no relationship between these two concepts. The only similarity is on the name that they were assigned, which is 'dark'. Probably because it is a word that refers to something that we cannot see or understand ... Finally, we can see now and it is clear that you all must have noticed that I wish to relate this dark energy that enters and comes in, with the black energy that escapes and goes out. This means that the energy of the black holes did not gone away or simply disappeared inside an endless an infinite tunnel!

All the chaos created by these complex structures, all of the entire monopoly of high entropy generated in these complex systems, the disintegration of matter in Black Holes is recovered and recycled in the form of Dark Energy by the Universe. Again, in the Universe, nothing is lost … everything is transformed!

Chapter XXII

HOW MANY DIMENSIONS?

"Beyond the stars inhabits other worlds."
- Einstein -

By postulating the entrance of this Dark Energy in our Universe, there is, of course, another question that immediately arises in my mind:

How many are the dimensions that surround us?

There will not be three, for sure. At least, there is one more. If there are others, I will simple not know!

The entrance of this uniform dark energy in our Universe should be postulated by a Fifth Dimension, ubiquitous, and always present, that involves us from all directions.

If the Universe is becoming each time bigger and more and more quickly larger, it is because there is something on the 'outside' that can constantly come in through this fifth dimension! This dimension is the hidden entrance of this energy, responsible for expanding the Universe at all points of space evenly.

We can even say that it is a fifth essence that enters through a fifth dimension. A Magic Energy, which leaves out through one door but enters through all!

Unless someone has a better idea, I am obliged to introduce this new concept, absolutely different and exotic, yet for us a little bit abstract and esoteric!

This mantle which involves us must hide many secrets ... the 'Bubbles of Oil' of the hyper-space!

The idea of living in an imaginary Universe with more dimensions is not new. Many theories have tried to explore these concepts. The Theory of Strings was one of them. For this theory to be viable, our Universe should have more than three dimensions, at least ten. In this Universe, the extra dimensions would not be visible, or they would not be perceived by us, in that case, these dimensions should be extremely small, so that, they were consider to be rolled. This rolled and small size dimensions would be much more difficult to detect than the larger size and extended dimensions, which are evident, the so familiar three spatial dimensions.

But our Universe may very well have much more dimensions than what it seems at first sight. And a new dimension will be, therefore, a new direction in space and time!

The multi-dimensions of the Strings Theory had its initial inspiration from the theory of two mathematicians: Kaluza and Klein.

In 1919, Kaluza sent a paper to Einstein with an explosive suggestion. Proposing that the space tissue could have more than the three common dimensions! Beyond those dimensions which we all know and that are provided by our physical senses and by our perceptions, there would be in our space tissue a fourth dimension!

If our Universe had a total of five dimensions, four for space and one of time, it would be possible to get a combination of unification between the Theory of General Relativity and the Electromagnetic Theory of Maxwell in a single common formalism.

Thinking that there may be more dimensions out there might be something with a peculiar sense, more or less bizarre. After all, what is the meaning of this new dimension?

Our three known dimensions are defined by the three possible directions of movement allowed in a spatial area, which are what we usually call of: left-right dimension, front-back dimension, and up-down dimension. These are the three possible spatial movements, accompanied always by a dimension of time, which is, the future-past dimension.

A new dimension implies and provides the existence of a direction independent of all the others, therefore, it can be consider as a new movement, a different way of experiencing and crossing the space and time!

But even if the Universe contains an extra spatial dimension, such a direction will reflect a physical concept rather difficult to conceive and to understand in our intellect. There are certain concepts which can only be perceived through mental abstractions.

On the other side, we cannot simply deny the existence of other possible dimensions. The concepts between what we believe that the world is and the concepts of what they really are, it will always be very speculative and controversial, subject of numerous descriptions and interpretations.

There is a difference between what we can give as a definition of real and what is the exact essence of reality.

Wised would be the modesty that would advocate a view more undefined rather than defined about the dimensions of the Universe. Our senses only serve to excite the reason, to provide us the good sense, to indicate and to testify, but they cannot see and know everything. The truth does not come from the physical senses, except a small part.

Until the appearance of the Theory of Relativity it seems out of the question that the Universe which we live in had more than three dimensions. But with this new theory, this old concept of three spatial dimensions had to be reviewed and it was establish and considered that the new space-time consists of only four unique dimensions, since time can be converted into a spatial component and vice versa.

Therefore, the geometry of Restricted Relativity applied to the old space-time is no longer Euclidean but instead of it, is now consider as Minkowskian; and in General Relativity, the geometry is no longer to be considered as Minkowskian, but Riemannian. The Math reveals itself, applied to new geometries and new dimensions.

Before the suggestion of Kaluza it was assumed that Gravity and Electromagnetism were two distinct forces, unrelated, with no relationship between each other, and that there was not a single chance of connection between them. But the creativity of Kaluza, imagining the hypothesis of a Universe with one extra spatial dimension, was a remarkable event, since it suggests, for the very first time, that there might exist a deep relationship between Gravity and Electromagnetism. That these two forces have geometric properties and that this property, the extra dimension, joins them together; making possible the missing relation, associating and uniting them, undoubtedly, to the wrinkles of the fabric of space-time. Naturally, everything that defines the space and time must be attached to it, as such, there should be a unification between Gravity and Electromagnetism.

The strange theory of Theodor Kaluza and Oscar Klein led to interesting results, if the space-time post by Einstein and Minkowski is added a fifth dimension, then, using their own field equations of theory of relativity, it is shown that the electromagnetic phenomena can be interpreted as having a geometric origin. In other words, the electromagnetic field and also the gravitational field are both geometricable! This means that these two concepts are not so distinct as they seem, that there is a possibility of unification, that these concepts cannot be attached with predefined and exact properties, but instead of it, the main properties of these fields should be classificate as undefined and adaptable to the very structure of space and time, being both very versatile and dynamic.

Or more simply, if the gravitational structure of an object provides a relative space and time field, the same gravitational structure is composed by electromagnetic forces which share the same properties of relativity. What it is obtained from here is that: both Gravity and Electromagnetism are geometricable are relative!

Under the hypothesis of a subtle extra spatial dimension, Kaluza has proceed with his work, undertaken this mathematical analysis in which at the relativity equations is added a new dimension and came to a set of new equations. After a study of the resulting of these new equations, corresponding to the addition of a new dimension, Kaluza understood that these new equations were not other than the Maxwell equations to describe the Electromagnetic Force!

This remarkable possibility, although it was an extremely beautiful idea, has been reflected in several subsequent detailed studies, but which were always in conflict and incompatible with the experimental data. It seemed that there was no way to confirm the existence of this new dimension.

However, this concept has continued to inspire new physical theories, namely the M Theory, or Theory of Strings. But ten or eleven dimensions are much more difficult to conceive than the five dimensions of Kaluza-Klein.

We also have a new theory emerging in the XX century, developed by the physicists Lisa Randall and Raman Sundrum, simply known as the Randall-Sundrum model. This theory also believes that we live on a hypersurface consisting of a space-time of five dimensions. It should be noted that this theory also uses all of the mathematical formalism developed by Einstein in his theory of General Relativity, only changing the dimensionality, which is five.

If it is shown that this Dark Energy comes to us through a fifth dimension, then, there still maybe some time to reconsider a five dimensions theory for our Universe and to demonstrate, experimentally, the discovery of this hidden dimension.

I remember a phrase I once read "Physics is first invented and then discovered. The hidden dimensions have been invented; they are only waiting to be discovered." - Carlos Romero -.

In a heuristic and intuitive way I would say that this hidden dimension will not have to be necessarily smaller, rolled and coiled; or larger and extended. It's not very likely that we will be able to specify the status of this new dimension, this one, it will simply be an involving dimension, ubiquitous and omnipresent.

Maybe just this concept of a new dimension, could be capable to explain the transfer and entry of this Dark Energy in our Universe and only perhaps this new multidimensional space could contain and include the very own limits of space itself, the topography of the Universe, the frontiers of the Cosmos.

If we conceive a horizon, a limit, a boundary for our Universe, we can always ask our self about what can exist beyond that frontier?

A Cosmic Abysm?

The question of the Topology of the Universe has shown a growing interest in Cosmology and raised several other possibilities and approaches. Some of these assumptions include a theoretical natural basis; others are focus in a pure mathematical and geometrical concept. Essentially what is asked is whether if the topology of the Universe is infinite or whether if it has a specific geometry and shape.

It will not be very easy to conceive a geometric model that includes the boundaries of space and time for our Universe. Although it is more convenient for astronomers to work with two dimensional flat surfaces, the real surface of the Universe is much more complex and it is a property that we have not yet been able to measure. If we conceive a three-dimensional geometry, such as a spherical surface, this will result in a Universe with a curve surface, coiled on itself, which would mean that if we could walk continuously in the same direction we would turn back to the same point.

But even if we could conceive the correct description of this space, the Geometry of the Cosmos, still so we would be limitating the space of the Universe in space. This means that we are imposing the limit of space in space!? ... which will always leads us to a paradox that we can never define where the end of space is!

Is it then possible that space could be infinite?

I do not wish to annoy you but, doesn't it seem to you that any specific limit or any type of topology that we can consider, it necessarily limitates space in space itself? So then, how many are the completely spatial dimensions of the Universe?

Is it possible that we exist in a Universe with infinite dimension?

The 'Problem of the Infinite' it is always an uncomfortable issue and a very inconvenient one!

But these pernicious infinites, challengers of our reasoning, are present and persistent in several points of the History of the Universe.

Chapter XXIII

ORIGIN AND FATE OF THE UNIVERSE

"At the entrance of the gates of the Temple of Science
It is written the words: 'You must have faith'."
- Max Planck -

The problem of the infinite is directly related to the origin and the fate of the Universe.

Both Math and Physics cannot deal very well with infinite numbers, but these sciences predict that our Universe was originated in a Big Bang full of infinity numbers! Starting in density, passing through temperature and ending in the curvature of space-time at the moment of the Big Bang, all of these properties are infinite in the field of Physics.

When a Physics theory tries to address these issues it faces a major obstacle, the equations result in infinite and in uncertain parameters, which means, in practice, that we cannot develop a consistent theory to accurately describe the precise moment of the Big Bang. If we cannot make a correct interpretation or give a concrete physical meaning to a set of infinite values, we cannot correctly describe and predict the process involved, in this case, contrary to what people regularly think, the Big Bang is not a theory that describes the initial explosion which originated our Universe, the Big Bang is a theory that only describes the result of the explosion, from a certain moment in time. As such, it is not a complete theory, because it does not describe the exact moment of the explosion.

These consequences arise from a very simple detail, which come from the fact that physicist considered that the entire Universe was originated in a spatial and temporal point with zero radius, and that this so compressed region is equal to a very compact dimension: a null or zero dimension.

What is the meaning of a zero dimension?! What possible process is intended to be described in a null space-time dimension?!

Isn't it clear that we cannot obtain from these conditions any consistent or any coherent physical theory?

But if physicists do can imagine physical processes that result from a zero space--time dimension, then surely, it is because they have a huge capacity of imagination, far better than mine!

There is some speculation on which our Universe has existed for an infinite time before the Big Bang had occurred, under the form of a floating empty region, before it has started the phase of expansion.

This seems to me an interesting approach. And in its theoretical formulation new phenomena begin to occur, peculiar phenomena which are not yet completely understood.

If we turn back in time to the initial singularity, we reach to a point where the theory of Gravity is no longer valid, and that is where new phenomena begin to occur. For example, if we turn back enough time, until a time before the Big Bang had occur, the calculations show us a Universe that is actually in contraction, until it reaches to a point that allows the phase transition, the point which the contraction process ended and began the expansion movement that we observe today. This moment has happened when the Universe reached a minimum size, but finite. Even more amazing is if we try to go back further in time, the calculations show us a Universe that is growing, much more extended and colder ... ad infinitum.

This theoretical speculation can lead us to suggest that there was not a proper beginning of the space and time when the Big Bang occurred, but that this exact time is only a state of transition from a duration of time much more vast.

Using this new concept, we can use our imagination and turn back in time until the moment of the initial singularity, a time very close to the Big Bang; or if you prefer we can also move forward in time towards the total contraction of the Universe, the moment of the Big Crunch.

Remember that the theory of Gravity is no longer valid from a certain value of density and temperature, because under these conditions we obtain a new type of matter, the Degenerated Matter, and according to what we saw on the new model of Gravity, the nature of Gravity is not quantum but atomic.

If this form of matter is isolated from a gravitational field, this type of matter is unable to produce its own gravity, because, as we have seen before, this attractive field is only present because the magnetic moment is distributed and transferred by the electromagnetic field. And if there are no electrons associated with atoms, or photons, there is no electromagnetic field, therefore, there isn't also any gravitational attraction.

This allows us to immediately eliminate the infinite gravities and the singularities addressed to these points. In this case, all we have to consider is that the intensity of Gravity, in a shorter radius increases and becomes each moment stronger, however, it will not assume an infinite value, but tend to a limit. From this limit the gravitational force ceases to exist and there is a phase transition, which defines the point where the uncontrolled contraction ends. At this precise moment, the Universe reaches its minimum but finite size, and the forces and fields of interaction that still exist in this limits of matter are monstrously strong, however, Nature has got a limit and so, as soon as Nature achieves this limit it will begin a new phase of uncontrolled expansion.

Right now, you may all be wondering about what is that limit of Nature.

Well, it will be a little as if Nature had reached the maximum stress and tension of its fundamental 'strings'. And if a force of tension has got a specific direction and tends to a limit, once this limit is exceeded, it breaks. This rupture is a bit like an elastic that is broken. This reaction occurs in the opposite direction, and the enormous force of tension that has been accumulated is reattributed in order to give rise to an enormous force of expansion. This expanded energy has a new form, it does not have natural gravitational forces, the forces that control this new force are naturally repulsive, this energy is already known: it's the dark energy.

And what is it that accumulates this tension? What are these fundamental 'strings'?

In this state of matter, the last force that will resist is the weak force produced by neutrons, beyond this limit we must considered another type of force, much stronger and even more fundamental, the strength force between quarks!

And when there is a force, there is an involvement field; and where there is a field there are lines of force which define it; and as we know, these lines of force never get cross. Nature has a limit, it reaches a point where these lines of force are becoming each moment more intense because they are closer to each other, becoming each moment even more closer and closer and stronger, but there is an absolute interdiction on the Laws of Nature, which is that these lines can never cross, because in Nature nothing can get touched!

Achieving the Constant of Repulsion is, therefore, inevitable!

The intensity of these lines of force results from the response of the quarks to the values of energy involved. We can remember that the quark particles are stable when confined in the nucleus. The Principle of Asymptotic Freedom establishes that the confinement of quarks leads to a state of minimum energy and we know through experience that the more energy we provide to these particles, the more resistive they become, thus means, more tension they accumulate. What would be a practical way to obtain a constant of repulsion, by the ionization of the nucleus, which would only be possible when we obtain these huge values of energy. The rupture of this energy of connection between quarks has enough power to unchain a huge and colossal explosion! A new Big Bang …

And so we would have a birth of a New Universe!

A Phoenix Universe reborn from the ashes!

The fate of our Universe is, in some way, already pre-destined. Probably we will only have two possible scenarios:

Or the 'gravitational force' responsible for the attraction of matter exceeds the repulsive force of expansion and the Universe is bound to shrink and compact toward a Big Crunch, becoming a region increasingly dense and increasingly hot … a cosmic furnace.

Or, the repulsive energy overcomes the attractive force - and this seems to me the most likely scenario - impelling the Universe to an endless expansion, making it an empty space, increasingly cold and completely inert ... a cosmic graveyard.

If in the first hypothesis there is an endless increase of Kinetic Energy due to the constant raise of intensity of temperature; in the second hypothesis, the opposite occurs, and in this case the kinetic energy of particles is becoming smaller, their movement becomes slower, as the temperature goes down towards the absolute zero, -273 Kelvin, the kinetic energy of particles practically cease, but on the other hand there is always a huge increase of Potential Energy.

Summarizing, if in the first case the Kinetic Energy is headed towards a maximum limit; in the second case the Potential Energy is guided towards its maximum value.

In both case, we have to consider the Principle of Energy Conservation. We know that the Mechanical Energy of the Universe, the sum of these two energies, must be kept, so that, the total mechanical energy of the Universe is always a constant.

In the first case, we have already explored the extremes conditions of a Big Crunch and that its development results in the birth of a new Universe.

For the second hypothesis the conditions are reversed, but the final destination is always the same, let's see how:

There is nothing worse than reaching the limits of the Potential Energy, which I would like to call it as a False Potential. We will recall here a practical example, a typical case of a lake exposed to harsh and sever temperatures of winter.

The phenomenon of overfreezing means that water can remain liquid even at temperatures below zero and under its freezing point. It is indeed possible to overfreez extremely pure liquid water and keep it fluid at a temperature of - 30 º Celsius, a negative temperature.

If we approach to a lake, still in liquid state, but which is in a status of overfreezing, it will be enough just a very small touch, the most tinny contribution of energy, for the lake to start freezing and unchain a whole explosive proliferation of ice crystals, bringing the lake to congeal almost automatically! I advise you never to touch on something that has acquired the False Potential.

Although the lake appears to be liquid and stable, in fact, this apparent stability is the most pure of the illusions. This overfreeze liquid is quite unstable and it is in a very precarious state of balance and equilibrium, and at the precise moment that we touch the lake, all of the Potential Energy accumulated is released.

Many properties of matter are changed when overfreezed, one of them is Gravity. Not that this property has been changed, the only restriction is that this force does not have the same possibility or chance to manifest itself.

Just like the attribution of Kinetic Energy in the form of heat produces an increases the gravitational attraction, as we saw in the experiment of Cavendish, the reverse process can also occur. Thus means that, a reduction of the Kinetic Energy of particles, can be achieve by exposing them to lower and negative temperatures, therefore, this process also leads to a decrease of the emission of gravitational energy.

When the Universe reaches the absolute zero we can say that the kinetic activity of the particles ceases almost completely and that Gravity also freezes. The Universe is reduced to its minimum temperature, its minimum density, it achieves the False Potential.

In this Universe filled mostly by vacuum and where there are no forces acting, Gravity forces or motion forces, the only new form of energy that may appear in this static Universe is the one which is naturally produced by a repulsive vacuum. But once that limit is reached, and vacuum acts contributing with its repulsive energy, so soon as it enters the smallest amount of energy in this Universe highly unstable, the high Potential Energy accumulated unchain a huge Cosmic Tsunami!

This small disturbance of the vacuum releases all the accumulated Potential Energy. A huge wave of energy that comes and succumbs, leading to the collapse of the entire Universe!

Fusions and fissions of Universes should occur regularly in the hyperspace, allowing the contribution of primary raw materials, new cosmic fluid, which will always be obtained through the Law of Conservation of Energy, so that it can always occur new creations of new Universes.

This Essential Fluid of the Universe, the dark energy, might resurrect the concept of ether, in which this ether will be related with an Eternal Fluid. From both possibilities, we will always obtain an eternal Universe, with not beginning or end, a hyperspace of Oscillatory Universes, a Phoenix Multiuniverse!

In any of the possible destinations for our Universe there is no infinite, Nature has always a limit. The contraction does not occur continuously and the expansion does not run forever.

The only infinite that Nature allows, which is eternal and continuous, is the energy conservation.

The energy cannot be created nor destroyed, so that the total energy of this Universe is a global constant. The soul of our own Universe is always conserved, because energy can naturally flow but it cannot disappear.

The testimony of the fate of the Universe leads to an Equation of Continuity, in which the creation occurs in a continuous space-time process, timeless and infinite. Because time preserves its own duration, from the infinity to the infinity, from the eternity to eternity.

And so we can consider us as part of a Universe of infinite life, which never began and will never end. In a way, this is a Universe that always existed and will always exist!

"I say that the whole Universe is infinite because it has no end, no limit, or surface. I say that the Universe is not quite all infinite because from each part we can take each one of them is finite and the countless worlds it contains each one is also finite. "

"It is not the physical senses that perceives the infinite, it is not from sense that is achieved that conclusion (...) it is the intellect that is responsible to judge and give reason of all absence things, separated from us by the distance of time and by the interval of space. " .

"It would not be less bad if the whole space was not full and complete. And as a consequence the whole Universe would have an infinite dimension, and there would be innumerable worlds. The intelligible extension is eternal, necessary and infinite. ". - Filóteo, character in 'About the infinity of the sky' of Giordano Bruno-.

The infinite is the most mysterious number of Mathematics, simultaneously absent and present, precise and vague, full and incomplete, perfect and imperfect ... there are those who love it and some who avoids it!

Nevertheless, the mathematicians have no major problems in using this number in their calculations: starting by considering Series as a sum of an infinite number of successive terms; an Integral as a sum of an infinite number of infinitesimal quantities; and a Derived as a ratio of quantities infinitely small.

In parallel, Physics also uses this number in order to refer major events of the Universe: A Black Hole is an infinite in space and time; the Big Bang is an infinite in space and time; and the Vacuum's Energy is also infinite in space and time!

What is the true meaning of this infinite number?

Using this number does not seem to cause many mathematically problems, but when it comes to give it a concrete physical meaning, that is when the confusion is install.

According to what we saw previously, two of the three infinites are no longer an epistemological problem, and physicist may feel a little bit relieved, because we can remove the analogy of infinite in Black Holes and the infinite of the Big Bang.

However, not to much relieved, because this reality does not extend to the Vacuum Energy. The vacuum, or the absolute emptiness, is not synonymous of nothing, it would be more correct to consider it as a mode of hibernation of matter. Simply because we just have to apply a minimum amount of energy on vacuum to stimulate the appearance of matter.

To demonstrate the real existence of these virtual particles of matter in the empty vacuum, physicist have provided a practical experience. By passing a strong electric current through a vacuum chamber, it is possible to see and assist at the creation of real particles, perfectly observable!

The energy of the vacuum cannot be associated with nothing, it is instead a level of oscillation and transition between a state in which there is nothing and a state in where there is something. The vacuum cannot be associated with emptiness, because anything can come out and be born from it!

The emptiness is a latent state of matter, and the vacuum is the floating state of that potential in which virtual particles are constantly vacillating in and out of existence.

These mysterious forces of vacuum, the infinite possible fluctuations of the quantum vacuum, are the properties of the amazing Energy of Nothing. What this energy translates is the uncertainty, that even in a small volume of vacuum there is no possibility of knowing the amount of energy that it contains!

When we wish to make a rigorous interpretation of what this means, we are perplexed by the paradox that it represents.

In practice, when this experience is conducted, all particles present in that region of space are removed in order to get an empty space and we think, according to a logical reasoning that by reducing the amount of matter present we are also reducing the existing energy.

But what is concluded is that at the Zero Point Energy the number of virtual particles can appear in an infinite quantity.

The paradox is: this means that the Zero Point Energy is not zero, it is unlimited! We can proceed our thought and say that the power of Zero is Infinite!

And this conclusions can be deduced by the very own equations of Quantum Mechanics, but most scientists completely ignore it. They simply pretend that the energy at point zero is zero, although they know it's infinite.

Infinites and zeros are two curious numbers. There are even those who say that zero is the closest number to infinity. So, are these two numbers so distinct and different from each other?

But it seems that we cannot deal very well with infinites and zeros. Can we simply exclude them from our equations? What do these defendants have to say about the Laws of the Universe?

These two numbers takes us back into the problem of the origin and fate of the Universe. The Problem of the Origin it only makes sense if we say that all began with the birth of the Universe: Time; Space; Matter... For a common sense that requires that there has been a beginning, this implies accepting the existence of zero '0'.

Moreover, we can shift the problem for the common sense which believes that everything has always existed in the Universe: Time; Space; Mater ... This means accepting the existence of an infinite '∞'.

A number almost complements the other because, in fact, they are very similar. They are so similar that we could speculate that origin and infinity coincide and coexist simultaneously ... the problems of infinites and zeros are eternally challenging the framework of logic.

When an equation has an infinite, the physicists usually assume that there is something wrong, immediately admitting that the infinity has no physical meaning. How the physicist Richard Feynman would say:

"The problem is that when we try to bring the calculations to a zero distance, the equation explodes up in our face and gives us answers without any sense, things like infinity. This has caused immense inconvenience when the Electrodynamics Quantum Theory was developed. People obtain an infinity in each problem which they were trying to calculate. "

Infinites and Zeros have always been present behind the greatest puzzles of Physics:

The infinite density generated by the gravity of a black hole is a division by zero in the equation of general relativity;

The infinite energy of vacuum is a division by zero in mathematics of quantum theory;

The creation of the Big Bang from the nothing is a division by zero in both theories.

The illogical situations and the indeterminations, arise every time we try to make calculations with divisions by zero. Apparently, dividing by zero destroys the coherence of the mathematical and logical framework.

However, instead of trying to understand what is the meaning and true expression of these solutions, the physicists succeeded in mastering the solution of the problem choosing another way, by passing it. In order to achieve and obtain the connection with logic, scientists simply banished 'zero' from the equations of Cosmos!

This process of elimination of zeros is called Renormalization. "It is what I would call as a kind of crazy process." wrote Richard Feynman, despite of having won the Nobel Prize for having developed and perfecting the art of Renormalization.

This process is extremely convenient and impose to physicist, however, not all agree with it. But, it is the only way to make the infinite to disappear, through the magical art of renormalization.

In practice calculations, it is not made all the way until the zero distance. Calculations stops near the zero, at a short distance more or less arbitrary.

Technically it is introduced very small measures of distance 'l', time 't' and mass 'm', with which it is now possible to perform calculations with having infinite solutions. These parameters are the so known measures of Planck:

$$l_{PL} = \sqrt{(\hbar\, G\, /\, c^3)} \approx 10^{-35}\ m$$

$$t_{PL} = \sqrt{(\hbar\, G\, /\, c^5)} \approx 10^{-43}\ s$$

$$m_{PL} = \sqrt{(\hbar\, c\, /\, G)} \approx 10^{-8}\ Kg$$

The initial success of the String Theory came from the eliminations of the zero in the equations of the Universe. Considering that, there is no distance or time zero. And so, all calculations were solved, especially all the problems of infinity. This, incidental changing, actually solves some problems; indeed, it makes disappear the infinity of black holes and of the Big Bang ... but, unfortunately, physicist did not realize the reason why it solves these problems.

I would say that the Planck's constants are a myth ... convenience constants...

Currently, it is in the process of Renormalization that relies all the foundations of modern Physics. I would say that these foundations are fragile, once they don't recognize the potential of these two numbers.

Zero and infinite, associated with energy, space and time are still being seen as a taboo in Physics. Throughout all History of Science, infinites and zeros were always found immediately guilty. But, what if these defendants are telling us the truth?

If they are innocent, we just have to convince ourselves that we are not prepared to assume the veracity of science that conglobates and contains these two numbers.

The mathematicians, who invented the math and numbers, the infinite and the zero, and the so many dimensions, work on this science but don't believe in it.

The richness and fullness of the Universe is in its own individuality and uniqueness. The fundamental reality of this Universe contains absolutely all the time, all the space and all the energy.

The infinite is a number that contains many numbers, which grows up with no limit. The Laws of Science, those, they do contain many numbers, but these are the Laws of Nature.

If scientists do have faith, they will understand the Laws of the Universe!

Chapter XXIV

Quantum theory of Gravity

3. UNIFIED FORCES

"It is at night that it's beautiful to have faith in light."
- Edmond Rostand -

Unifying Forces! Where is the formula for a Unifying Force?

As we all know, all physicists do like formulas!

Oh yes ... the formula! The truth is, I must confess, I did not found the formula. This formula has already been discovered by the contemporary physicists, but they still do not know that they have already found it!

What all physicists do wish to achieve is the official unification of the Quantum Theory with a Gravity Theory.

It is practically impossible to marry the classical Theory of Gravity with a Quantum Theory of everything else ... how many times did I have already heard this sentence! It seems impossible to join these two theories together. Perhaps that is so because, maybe, they do not wish to get marry! If that is their will, let them stay single but happy!

The Standard Model works very well for all the other three forces of Nature, except for the Force of Gravity. The reason why this problem remains is because we are still looking at these four forces as having the same origin and nature. Considering these four forces as being all original of the Cosmos, but what we have here, if we look more closely, is a clear problem of perspective!

It is true that we see four forces, but the subtlety is to realize that not all of them are original forces of the Cosmos. One of them is a secondary force, only three of them are really quantum forces, because Gravity, as I said before but I will say again, is not an original force of the Cosmos but a consequence, a side effect. Of which can be concluded that what we call Gravity is an atomic force and so it is not a quantum force, as such, it cannot be included in the Standard Model.

This debate which aims to unify the three pattern forces of Nature: the Weak Force; Strong Force; and Electromagnetic Force with a fourth one: the inconvenient Gravity, is now eliminated!

This scientific speech which has continued and persist for years; this scientific saga which physicists have carried for a long time is then resolved.

And maybe the search for a Unified Formula is just to mislead. Perhaps it does not exist such a formula for this unification.

The only unification which I can present is the one of the current Standard Model. But this unification which unites three Forces: Strong; Weak and Electromagnetic, physicist already have it ... or almost. The unification is still pending in the Strong Force and the Electroweak Force.

Making a vain analogy, I would say that physicists have first discovered the developments of Maxwell Equations before they even find the formula of the Electric Field. Which is, actually, much more difficult to accomplish. But, somehow, they have succeeded. The two formulas of this unification are very complicated, there is still missing a final rearrangement and a simplification.

The theory of the Strong Force is studied by the Quantum Chromodynamics, short known as QCD, whose presentation in the language of Physics is:

213

STRONG FORCE FORMULA

$$L_{QCD} = -\tfrac{1}{4}\, F_a{}^{\mu\nu}\, F_{\mu\nu}{}^a + \sum_f \Psi_f\, (\, i\partial - M + g_s A^a T_a\,)\, \Psi_f$$

$$F_{\mu\nu}{}^a = \partial_\mu A_\nu{}^a - \partial_\nu A_\mu{}^a + g_s + f_{bc}{}^a A_\mu{}^b A_\nu{}^c$$

The theory of Electroweak Force is studied by the Quantum Electrodynamics, short known as QED, whose formula is:

ELECTROWEAK FORCE FORMULA

$$L_{E-W} = Lg + L_f + L_H + L_m$$

$$L_g = -\tfrac{1}{4}\, G_a{}^{\mu\nu}\, G_{\mu\nu}{}^a - \tfrac{1}{4}\, B^{\mu\nu} B_{\mu\nu}$$

$$L_f = \sum_i \Psi_{Li}\, (\, i\partial + g' W^a t_{a} + g\, B\, y\,)\, \Psi_{Li} + \sum \Psi_{Ri}\, (\, i\partial + g\, B\, y\,)\, \Psi_{Ri}$$

$$L_H = -\,(\, D_\nu \phi\,)\,{}^\dagger (\, D^\nu \phi\,) - \mu^2\,(\, \phi^\dagger \phi\,) + \lambda\,(\, \phi^\dagger \phi\,)^2$$

$$L_m = -\sum_{i,j}\,(\, c_{ij}\, \Psi_{Li}\, \phi\, \Psi'_{Rj}\,)$$

As you can see, the expressions which describe these interactions are horribly complicated, but they do show a mathematic consistency which is in full agreement with the experimental predictions and, therefore, among its main protagonists, some were reward with the Nobel Prize.

But there is a possibility for a simplification...

First, the Strong Force is, with no doubt, the most complex force of Nature and the most difficult to simplify.

We know that the Strong Nuclear Force acts trough protons and neutrons inside the nucleus, keeping them together and preventing the electromagnetic repulsive force between protons to manifest. However, it would be more accurate to say that this force acts, not quite between protons and neutrons but rather between their most fundamental constituent particles, inside the level of its internal constitution: the quarks. The Strong Force joins protons together but also joins neutrons inside the nucleus, because it's a force between quarks. In a way, we can ignore that there are protons and neutrons in the nucleus and consider that within this system there are only quarks.

Inside this system there is a significant degree of complexity that arises when ever we want to study the physical behavior of this force.

While in the electromagnetic force we only have to consider one photon as the mediator particle of this force, with the strong force we must considered eight gluons particles mediating this strong interaction. These different categories of gluons can occur randomly and with an additional feature. Again, making an analogy with the electromagnetic force, the photons are electrically neutral particles, and therefore they do not interact with each other. But with the Strong Force that does not happened. We can say that these gluon particles can also feel the strong force, and therefore they interact with each other. Under these conditions, the treatment for this force is a little more demanding and exigent. And that is why it comes the complexity of the equations.

Without wishing to move towards the complex details of Quantum Chromo-dynamics, I can say that this is a science that has just started to make their first steps, because its deeper knowledge still escape and eludes our understanding.

Essentially, we can say that the fundamental particles of the whole matter are quarks and leptons. We have seen that leptons and quarks always form duplets and also that they react in a similar way to the Weak Force. However, leptons and quarks appear to behave quite differently to in the presence of the electromagnetic force and the strong force. Since the quarks feel the strong force and leptons don't. Moreover, once quarks have a fractional charge of + 2/3 or -1/3 and leptons (electrons and neutrinos) have a whole charge number of -1 or 0, the power of participating in a electromagnetic field varies from one to another but only in a quantitative manner, because the electromagnetic properties expressed are, in fact, identical. Both quarks and leptons feel the electromagnetic force.

There is a triangle that must be accomplish and that means that all forces have to work together, which suggests once again that these forces are not as distinct and independent as we think.

Until very recently it was thought that these forces had no relationship between them, since they act in a very different way and also with quite different intensities.

The power of intensity of these forces is an interesting measure which allows us to make a relationship between them and, as we know, the intensity of these forces differs from each other. Even the intensity of the forces varies, depending of the environmental energy, it is therefore not a constant value but a function of temperature.

In the cold Universe of today, these intensities are well defined and can be related through the Fine Structure constant.

The Fine Structure constant is a very special constant in Physics, and characterizes and describes the intensity and magnitude of the Electromagnetic Force. It is defined as:

α = 1 / 137. This alpha arises from the unification of three fundamental constants...

$$\alpha = \frac{e^2}{2\,\varepsilon_0\,h\,c} \approx \frac{1}{137}$$

... The electric charge, the permittivity, the Planck constant and the speed of light. These constants represent and combine different components from different areas of Physics:

c = relativistic component;
h = quantum component;
e = component of the electromagnetic interaction
ε_0= component related with the charged particles of vacuum.

This constant is actually extremely important. If our Universe is the way it is today, that only happens due the value of this precious fine structure constant. If, by any chance, this constant were modified, making it a little stronger or a little lower, if alpha had a different value, even minimum ... the essential characteristics of our Universe would no longer be the same and everything would change.

Disturbing this constant involves changing the energy levels of the atom; meaning that the values of the atomic energy connections would be different ... implicating a major modification in the whole atomic process! This would mean, in practice, that we would get a new Universe!

The coupling constant that defines the electromagnetic force is the reason to be of everything around us.

The Standard Model aspires to obtain a unification of the various coupling constants in a high energy environment.

The Grand Unification Theory requires that the natural forces and the particles on which they act, produce, in a high energies environment, a uniformity that is obscured at low energies. So that, it is briefly concluded that these forces act in a similar way in a remote past in the form of a single Unifying Force.

The introduction of this idea allows the high energy, physicists deduce that there is a similar behavior between the main forces of Nature when they are exposed to very high energies, in the order of 10^{15} GeV and, therefore, they consider that there was a primitive unification at a time when our Universe was still very hot and young.

The Unification presented is, as such, a function of the scales of energy involved. The theory provides a different evolution with distinct intensities for each force of Nature, according to the increase of temperature, as described in the following graph.

The development of this evolution seeks to identify a magic point in time, the point which would unify and change all Forces of Nature into one!

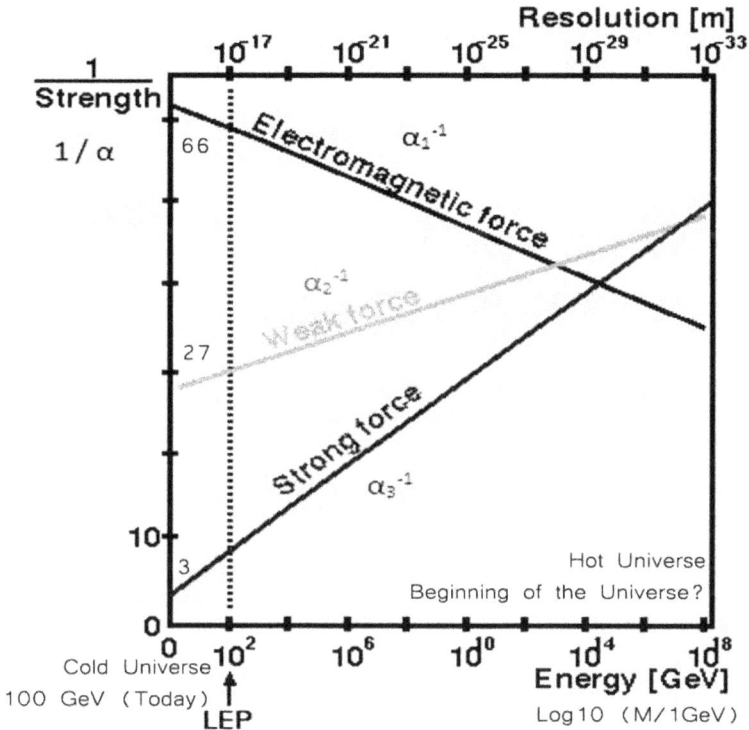

- Unification of coupling constants α -

The idea of unification does have some meaning. It is clear that Particle Accelerators show us new processes of mutation and disintegration of particles that can only happen in high energies levels. However, even agreeing that the intensity of forces converge into higher energies, how can we be so secure that their properties and characteristics also change and converge to a single unified force?

According to the current Model of Unification it is consider that there are several steps in the liberation process of the four Forces of the Universe. First it is considered that the four forces of Nature were join together in one single branch and that the spread branches of the forces happens in a partial and gradual process. With this process of ramification results that the Force of Gravity was the first force to get separated and to emerge from this Grand Unified Force, however, as we have seen before, it is exactly the opposite that happens. The Force of Gravity is the last to emerge and it is, therefore, the youngest force of the Universe!

And for the following ramifications it would be necessary to clarify a little better the reasons and justifications for this process.

In my point of view, there is no unification. There is, quite simply, evolution. And very hardly the forces would change their properties as if they were the same substance in a phase transition ... I do not think that it is reasonable to treat the Forces of Nature as liquid water, ice and water vapor ...

As we saw earlier, the justification is simple and it is withdraw from the cosmological evidence.

We can notice that all forces of Nature have a common characteristic, which is to unify, and as such the Electromagnetic Force, the Strong Force, and the Force of Gravity respect the symmetry. However, there is a force that does not share this characteristic of unification, because its main function runs exactly in the opposite direction. Its power is to disunify and it is, therefore, a force of imbalance and instability.

We can consider that the Weak Force is within the origin of all forces of Nature, once this is the only force which breaks the symmetry. As so, it is from this dissymmetry that evolves and emerges all the other three forces that surround us.

Thus, we can simplify the development and evolution of the forces of Nature, proposing a more balanced and symmetric branching:

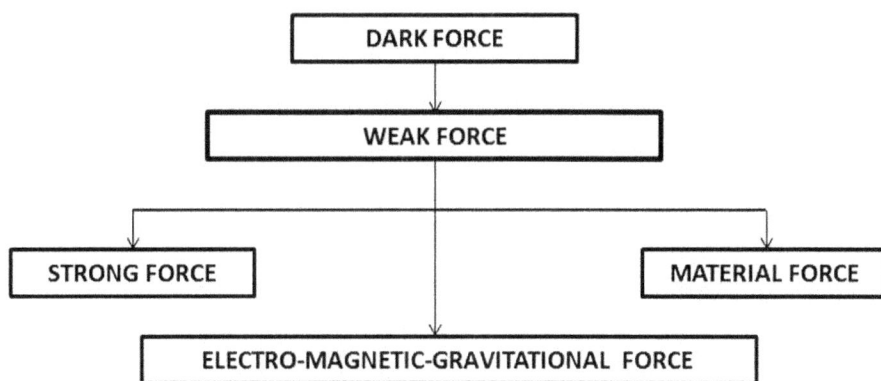

```
            ┌─────────────────────────┐
            │       DARK FORCE        │
            └─────────────────────────┘
                        │
                        ▼
        ┌─────────────────────────────────┐
        │          WEAK FORCE             │
        └─────────────────────────────────┘
          │               │             │
          ▼               │             ▼
┌──────────────────┐      │    ┌──────────────────┐
│  STRONG FORCE    │      │    │  MATERIAL FORCE  │
└──────────────────┘      │    └──────────────────┘
                          ▼
        ┌─────────────────────────────────────────┐
        │ ELECTRO-MAGNETIC-GRAVITATIONAL  FORCE    │
        └─────────────────────────────────────────┘
```

- Evolution of the Forces of Physics -

With this structure in mind, we can emphasize that it was not the temperature that has conditioned all these physical processes. The external factor of greater relevance which more conditioned these circumstances is according to a natural process of Evolution and its influence is predominant and crucial ... the external agent more effective in this process was ... the Time!

221

I must remind you that the Particles Accelerators cannot reproduce all forces ... and time is definitively ... a huge force!

Time is a great force but given its extraordinary subtlety and discreet nature of its activities, it's a force which nobody notices...

But this force acts constantly, cleverly, without being noted. It is an actor that hides its true nature and identity, by showing, or to be more accurate, by hiding behind their actions among things.

It is with its actions that we must count on, in order to draw a complete perspective of the Evolution of the Universe, it is with this action that we must conceived the whole History of the Cosmos.

Time is also a Force ... always forgotten ...

The simple structure, functional, harmonious and beautiful of our Universe is due to the complementarities and interactions of all these forces.

Returning to the theory of unification, it is possible to compare the magnitude of the forces with each other and thus obtain a relative magnitude.

The main relationship is made from the fine structure constant.

Usually the magnitude of the Electromagnetic Force is stipulated as having a value of 1/137 and this force is considered to be 137 times weaker than the Strong Force;

The Weak Force is by its turn 10^6 times weaker than the Strong Force;

And the Gravity Force is the weakest of all, in the order of 10^{40} times weaker than the Strong Force.

With these data we can establish a relationship, in which the more intense force is the Strong Force and we match it to the unit 1:

Strong Force = 1

Electromagnetic Force = 1 / 137

Weak Force = $1 / 10^6$

Gravity Force = $1 / 10^{40}$

We can also adjust these equations by multiplying the intensity of all forces by 137. And we began to define the coupling constant, not according to the Strong Force but in order of the Electromagnetic Force:

Strong Force = 137

Electromagnetic Force = 1

Weak Force = 137×10^{-6}

Gravity Force = 137×10^{-40}

With this definition, there are several relationships which can emerge from this structure, as for example:

$$F_E = 1 \times F_E$$
$$F_S = 137 \times F_E$$
$$F_W = 137 \times 10^{-6} \times F_E$$
$$F_G = 137 \times 10^{-40} \times F_E$$

Other equivalences can also be deducted. Transforming the equation and considering that $F_{EM} = 1 / 137$ x F_S, appear as:

UNIFIED FORCES:

$$F_E = \alpha \ F_S$$
$$F_S = \alpha^{-1} \ F_E$$
$$F_W = \alpha^{-1} \times 10^{-6} \ F_E$$
$$F_G = \alpha^{-1} \times 10^{-40} \ F_E$$

With the exception of the Material Force, about which we know very little, this is our alternative for the unified and simplified formulas of Electromagnetic Force, Strong Force, Weak Force and Gravitational Force. Thus avoiding to present these equations through extremely complicated developments.

Considering that we know the relationship of alpha 'α' as also the formula of the electric field F_E, it is possible to obtain a decomposition and resolution of these forces.

However, the theoretical physicists wish to facilitate the calculations and, therefore, they would like to obtain one single Unified Force ... but if the forces are exactly these ones, then, there is nothing else to simplify ... there is no Unifying Force!

At this point I am reminding me of the classification for the scientific system according to the physicist John Barrow: 'Every time a new idea is presented and notice in the scientific community it necessarily has to get through three steps: 1st It does not worth anything, we do not even want to hear about it; 2nd It is not wrong, but it probably has no relevance; 3th It is the greatest discovery of all time and we were the ones who found it first. And by then there will be many more candidates claiming the priority of the discovery. "

It is unfortunate that some scientists never deviate from what is already known, always choosing to follow the safer trail. They do not even risk crossing the berm, transgressing the trail, or even give any chance to an alternative theory.

It is their responsibility to expand the frontiers of Physics and open new horizons.

Clearing new trails and discovering something completely new is fascinating! What's more magic in Science is the ability to go and get lost in the jungle of the unknown.

"The most beautiful experience is the encounter with the unknown." - Einstein-.

Science is above all a human activity very rewarding. Perhaps the most purest of all in a world far from perfection ...

Only scientists feel very deeply what they do and only they feel real passionate in what they seek.

"In the Universe no one has more fun than us." - João Magueijo-.

Chapter XXV

FORMULA OF TIME

"Time does not exist. It is only a convention. "
- Jorge Luis Borges -

What is Time? What is Space?

We could begin by naively assuming that we do not know what Time is, that we do not know what Space is.

We only know what is a relationship of space, that space is a distance between two points, but we are not capable to say what is the concrete and absolute meaning of space itself.

We also cannot say what it is the precise definition of time, we only know what is a relationship of time, that time defines a duration between two events.

But defining objectively these two concepts becomes very difficult and it almost locks our thought every time we try to define the origin of time and the limits of space.

Once you invoke the idea of an origin or a beginning, and therefore, a start, this abstract idea immediately transcends us. Let us admit that the Universe did have a beginning. But considering that the beginning is a moment of existence preceded by a time before this beginning had started, so we can say that there was a time when our Universe did not exist, the time before the origin of the Universe; and in a way what we are considering is a time before time, so that we can define it as a pre-time or, let's say, an empty time. But, speaking about the beginning of time itself, even considering it surrounded by an empty time, a time when nothing sill existed, it means that we are still situating time in time ... we are continuing the same limitation of time ... in time. That is a vicious cycle!

Nobody understands how can we design a creation of time outside time itself. Possibly because that conception is not possible!

There isn't, by definition, a period before time, nor can exist. So, the question of what could exist at that time, before the beginning of time, has no meaning.

It is always dangerous to ask what could have existed before the birth of the Universe. If you believe that time didn´t exist in that period, what we are considering is that there was a time when time didn´t exist. Please do notice the paradox of this last sentence ... " a time when time didn't exist" ... so what we are saying is that there was a time when time didn't existed!

In fact what we are saying is that there was a pre-time, maybe different from our usual physical meaning of time, but still so, in practice, what we are doing is assigning a new concept for time, which in fact is far from answering our question, just transforms it and modifies it into another perspective.

If we assume that there was no time before time, then there could also exist no History to be situated it in that time before time. Concluding that if there was nothing to be situated in time there could not exist any Historical time ... as well it could not existed any historical time around, surrounding a time when time didn't exist, thus mean, the time when our History is included. What I intend to say is: Accepting our historical time around a time that didn't exist is equivalent to situate that time that never existed in time in time itself ... which is an absolute paradox!

Therefore, we must conclude that there is no possibility of conceiving the existence of time outside time, and as such, that time has always existed and that there never could have been a beginning of time. The

227

only possible and logical statement is that time is eternal, immortal and infinite!

Let us now move towards a more fundamental question which is:

Is it possible existing several moments of time ... at the same time!? Or even different types of time! Is time an absolute measure, effective, unchangeable, always equal and immutable in its form, always constant in all fields of the Universe and in all Universes, or is it possible that time could be a more variable and versatile concept?

If you look well, this Force of Time is the dominant force! Very subtle, but always present and active. This force reigns since the beginning of time and it was already present long time before all the other forces have emerged. I would say that is a kind of Supreme Force.

But nobody has ever heard about this Time Force! How come something so obvious to us can go unnoticed?

I understand that it is often when things are under the disguise of obviously that no one can see them.

What are the main features of time? There is nothing in known Physics which establish a law for the passage of time. Is this concept still too abstract?

We have already seen that it is not possible to conceive a beginning of the Universe outside of time. The same argument is valid if we conceived the end of time of the Universe ... so this means that our Universe can have no end! Even in this time when everything would have an end, even in this world where nothing is happening, a cold and dead Universe, static and inert, even then when nothing is moving, the Time itself, is the only thing that remains active and alive to continue to be and do its work.

THE TRAVEL IN TIME [Scientific Version]

Time it's the artisan of its own preservation, of the renovation of Present, of the continuity of time. Its true stop would mean, not only the immobilization and absence of movement of all things, but also the immediate interruption of Present, and with this, the disappearance of all that exists. When I say everything, I truly mean to say absolutely everything! What is difficult idea to conceive in our intellect. This instantaneous and complete annihilation would result in a true Apocalypse! In this apocalypse everything would have an end. The complete abolition of time what would mean the end of the 'end' itself, as such, there could not even ever existed the moment of the end. Because the moment of the end also follows the principle of causality, which in our case would mean that there would not have existed a cause to precede the end. And if anything leads to an end ... then, there could not exist an end at all! What I want to say more clearly is that the end of time is not conceivable. We would face another paradox.

There cannot exist a world without time, or time outside of time. So the time is consubstantial to the world and to time itself.

Thus means that time cannot be finite in both directions; time must be infinite in all directions. We cannot trust in a asymmetric model of time for our Universe, establishing a beginning of time in one side and an infinite or indefinite time at the end of the Universe.

Time is symmetric, constant and infinite in all directions and in all dimensions.

Let us try to decode a formula for time.

If time is everything at it is everywhere, absorbing everything that happens and even what does not happened ... where is the fundamental formula of the time?

How come that physicist aspires to build a Unified Theory without including the Force of Time? Time is also a force, a strong one, this is also a seed of our Universe. How can we imagine to build a Universe without time?!

First, it is possible to demonstrate, mathematically, that time does not exist; that time is simply an abstract concept.

When we say that time has no real existence, we are considering that time has got, simultaneously, a physical magnitude null and infinite. This Fundamental Force of Time can reach quite distinct values, since the value zero up to the infinity.

According to the deduction of António Saraiva, this mathematical demonstration is possible. Beginning with the assumption that zero is equal to infinity, it appears that:

$$0 = \infty$$

$$\log 0 = \log (+\infty)$$

$$-\infty = +\infty$$

$$\log (-\infty) = \log (+\infty)$$

$$\log (-1) + \log (+\infty) = +\infty$$

$$i.\pi + \infty = \infty$$

$$\infty = \infty$$

The conclusion is that zero is equal to infinity, and vice versa, that the infinity is equal to zero.

This theoretical formulation can be transported to a practical formulation, as follows: First, we can consider that time has no real concrete existence. In this case, we are assigning a zero value for time. What this means is that we can say that there is no physical and objective existence of a temporal duration, in practice, it means that we cannot usually say 'there goes that little bit of time'. Because time does not exist as a physical entity in space, we cannot find it with physical dimensions of an object in particular, because the natural time is not formed of three space dimensions. Time, as a relationship of duration between two events, past and future, has got a null physical existence, and that is the reason why we cannot find it in a physical way. More simply, time takes value number zero because time is not a physical object. And that is why time doesn't exist ...

Furthermore, we all feel the passage of time. But this time which we all perceived is a Present time, one single moment, it's a moment with no duration or dimension, it is a singularity. And so, therefore, Present time is infinite. This is when time takes its other side, the value of infinity.

Time assumes an ambivalent and multifaceted feature, simultaneously null and infinite, present and absent. So that we can say: in Time nothing is eternal and nothing is ephemeral!

If we wish to understand this concept by the nature of the physical perception of our senses, we will stay completely limited. This means that, if we cannot handle Time physically, maybe we can understand and treat this concept mathematically. But in order to proceed to a mathematical manipulation of time, we need to know a fundamental formula of time. What is the mechanism that encodes and decodes the constant temporal relationship? What its messenger, the particle of time?

So, let's see, does time affects all organisms and all objects simultaneously and at the same way?

A bacterium is scheduled to reproduce in a very precise rate of time. The timing of cell division of bacteria varies from species to species and it is also influenced by many external factors such as temperature. But in the best growth conditions of these unicellular organisms they can perform a cell division every twenty minutes. This means that a single bacterium can produce 280 trillion descendants in one day!

A radioactive mineral is programmed to disintegrate at a specific rate of time, designated as the half-life period. This decay period corresponds to the time required for half of the original nuclide to disintegrate into a more stable nucleus.

The trees of the family of conifers, called Sequoias, know that their life expectancy is estimated at an average of 3500 years of time. Indeed, the oldest tree in the world is 9500 years old; it's a common pine in Norway.

How come everything that exists knows how much time is passing by?

Possibly because all material bodies have their own internal biological clock, which keeps them tuned with this frequency of time. Only that, nobody knows what are those exact mechanisms which contains this enigmatic Force of Time. This vast empire is still waiting for an astute and courageous explorer to find the absolute domain and knowledge of the flow of time.

If we could discover the basic mechanisms of time, their most intimate secrets, we would have in our hands the greatest force that governs the entire Universe!

If time is what makes everything change at a constant rate, with the same speed for all beings and substances that have the same properties, so that nothing can remain equal to itself, we can say that Time is a Constant of Changing.

But if time is also what makes everything to continue, so, in a similar way, we could also consider it as being a Constant of Continuity. The changing and the impermanence require a timeless law of eternity and continuity. In an obvious sense, we can say that we are all travelers in time, because, even if we do nothing at all, we will still be inexorably dragged under by this force of time into the direction of the future. Thus, we can deduce that time ensure and guaranties that all matter remains informed about the passage of this force, without exceptions. There is always a pattern in time and a sintony which sensitizes all molecular substances and all chemical elements for the passage of this force.

And what is the main key, the fundamental constitution of biological and non-biological organisms? What is the fundamental constitution of matter?

Again we return to the Field and Energy! So that we can say that time is a form of energy, like so many others. Turning the redundancy, we can say that time is the energy that affects energy. Which means, quite clearly, that time also travels through space in a uniform, constant and continuous way, tracking the speed of light, because this temporal field also distributes at the speed of approximately 300.000 km / s. Because Time is a Field associated with Space.

Time is a field! And this temporal field is, in its essence, uniform and homogeneous. Of course, there may be individual cases in which the very structure of space is altered and distorted. In such cases, if time spreads through space, or rather, it's a prisoner of space, we can assist to different flows of time ... Because the expression of time can be changed ...

If the energy of time is dominant, always present even in the absence of all other forces, what is the field force which produces it? If the energy of time is constant, what is the field responsible for giving this consistency?

This Common Time, which we are surrounded, is, first of all, a particular time and it is not the Fundamental and Absolute Time.

Another interesting physical relationship, established by António Saraiva is: Considering a Hydrogen atom, we can determinate that an electron travels an orbit in a fundamental and curious pattern. In this fundamental movement the electrons velocity is minimal and its energy is also minimal. So that this movement follows a constant cycle with a well defined period.

More important than what numbers can show us, is the theoretical interpretation that arises from this. And this is the part we must highlight: that all atoms in the Universe follow a constant pattern. The dislocation of the electrons charge in an atom follows a well defined period. This repetitive motion is the same for all atoms. And these are the biological clocks of Nature.

The atoms themselves were designed and created from the beginning to respond to a specific flow of time. As such, if any object also have atoms in its constitution, all the entire structure of more complex groups of matter produced from them, from molecules to cells, to life itself, follows this same pattern of time.

Thus it is easy to conclude that it is the speed of the electrons in atoms that sets the course of time, the life-time of an atom.

If we could build a physical equipment capable of measuring and recording the number of orbits completed by an electron in an atom, it would be possible to verify that this number is a Constant. Except when this atom stops to be at rest and starts to move with speed, and then we would assist to the phenomenon of General Relativity and conclude that the number of orbits performed were clearly lower. The result of this process is reflected in a reduction of the flow of time.

To understand the subtlety of this force we must again remember about that valuable constant: The Fine Structure constant. This constant re-enter on stage in a more fundamental question of Physics, which is Time. In the annotations of A. Saraiva we can see the perimeter P, corresponding to one stable orbit, has an exact value as it is showed in the following equation:

$$P = 2\pi R_b = 137\lambda_e$$

Where:

R_b => Bohr radius = 5.292 x 10^{-11} m

λ_e => Compton wavelength = 2.426 x 10^{-12} m

The Bohr radius (or average radius of an atom), and the Compton wavelength (or the average wavelength of an electron), represent units of measure that allow us to determine the perimeter of the path made by the electron. This perimeter, as we can see, has got a very precise value, exactly 137λ_e. Similarly, we can also try to predict the average speed of the electron while it moves in this stable orbit, which also follows an interesting relationship, by the following equation:

$$v = c / 137$$

Again, the Fine Structure constant α = 1 / 137, seems to have an important role in the speed of this particle, or rather, in the speed with which follows Time itself ... and, interestingly, associated with the speed which Time is spread, the speed of light c. We can define that the velocity of time, or the Formula of Time, is given by:

$$v_t = c\ \alpha$$

235

Chapter XXVI

FORMULA OF THE COSMOS

"God wishes, the man dreams, and the work is born."
- Fernando Pessoa -

It should be good to begin by recalling that if Physics had a Theory of Constants, it would find one single constant, unique and universal. That means that, eventually and inevitably, it would be discovered the Constant of Nature, the Fundamental Constant!

There is a constant that dominates the Universe and that is our constant of reference, the constant that identifies our Cosmos.

Unlike the speed of light 'c'; the electrical charge of the electron 'e'; and the Planck's unit 'h'; this new constant is still unknown to all.

The deduction of this constant is quite simple to demonstrate.

The formula that I will present establishes a very interesting relationship, as follows:

$$a = Q.c^2$$

Substituting the values in this equality, the charge of the electron 'Q' and the speed of light 'c', is that:

$$a = 1{,}6 \times 10^{-19} \times (3 \times 10^8)^2$$

$$a = 0{,}0144 \quad C.m^2/s^2$$

Considering the Einstein formula for energy, it comes that:

$$E = m.c^2$$
$$c^2 = E/m$$

Relating to the above formula, is that:

$$a = Q.c^2$$
$$c^2 = a/Q$$

From where it can be deduced that:

$$c^2 = c^2$$
$$E/m = a/Q$$

Transforming and matching the two expressions it is obtained two fundamental equations of Physics:

$$Q.E = m.a$$

The formula of Electric Field (F=Q.E);
and the formula of Newton for Movement (F=m.a).

The next question is: what does it mean that a = 0.0144?!

If you notice carefully, there are three most fundamental equations of Physics, which can be practically reduced to the following:

$$F = m.a \ ... \ F = Q.E \ ... \ E = m.c^2$$

It is by the combination of these formulas that we obtained this precious constant:

$$F = m.a... \ F = Q.E \ ...$$
$$m.a = Q.E \ ...$$

$$E = m.a \ / \ Q... \qquad E = m.c^2 \ ...$$

$$m.a \ / \ Q = m. \ c^2 \ ...$$

$$a \ / \ Q = c^2 \ ... \ \mathbf{a = Q. \ c^2} \ ...$$

The value that is obtained is a Force value, a constant of a Fundamental Force present in our Universe.

What kind of force is this?

Only later I realized that this is a Gold number, the number of the Cosmos!

This force is associated with the movement of gold which absorbs our entire Universe, from the Astrophysics to the Microphysics.

As we have seen before, all bodies of the Universe seem to have a circular motion responsible for this delicate balance of Nature. This kind of movement is perfect and perpetual, just like a pendulum. It only needs that 'someone' provides the very first initial clicking...

The natural essence of matter itself, incorporates this fundamental quantity; which corresponds to the essential natural Flow of Energy that is always in constant motion ... everything is energy in motion.

This fundamental value is associated with this movement of rotation that is always present and involving us. We can associate it with an angular momentum of energy ... the energy that placed our Universe in motion.

In this closed system, our entire Universe has an angular momentum of energy eternally constant.

FORMULA OF THE COSMOS:

$$a = Q. c^2 = 0,0144$$

$$a = 0,0144$$

Chapter XXVII

FORMULA OF THE UNIFIED THEORY

"The number dominates the Universe."
- Pythagoras -
-

What if we did find the Final Theory of the Universe, the Great Unified Theory? ...

What would that discover mean? ... What would we do with it? ... What would that change?

Would we be absolutely certain that we did find the correct theory? ...

If we had in our hands a mathematically consistent theory, which could make the correct predictions of physical events, always in agreement with the observations, we could acquire a reasonable degree of confidence, we could think that maybe this is the true theory, we could fall into the illusion that we have conquered the absolute understanding of our Universe...

"However, if we discover a complete theory, which shell be understood within a few time, in its general concepts, by everyone and not just by some scientists. So all of us, philosophers, scientists and just ordinary people, will be able to take part in the discussion of knowing why we and the Universe exist. If we find the answer to this, it will be the ultimate triumph of Human reason, because, with the knowledge of this theory we will understand the mind of God. "- Stephen Hawking -.

Physics will be mature when it reaches a Unified Formula that fits within a T-Shirt. Something more or less of this type:

$$F = m \, a \; ... \; or \; ... \; E = m \, c^2$$

A Unified Formula that will be an authentic work of art and which will immortalize its author ... contained in one single constant that will also reveal the reason of all the other constants that surround us.

Understanding the origin of our constants is one of the big questions of Physics...

... This formula could be contained in one fundamental constant, as simple as...

FORMULA OF THE UNIFIED THEORY:

$$a = 2 \, \alpha$$

$$0,0144 = 2 \, . \; 1 \, / \, 137$$

$$0,0144 = 2 \, . \, 0,007$$

$$0,0144 = 0,0144$$

From this simple number follows the full complexity of our Universe:

$$a = Q \, c^2 = 2 \, \alpha$$

$$a = 2\alpha = 2 \cdot \frac{e^2}{2 \cdot \varepsilon_0 \cdot h \cdot c} = \frac{e^2}{\varepsilon_0 \cdot h \cdot c}$$

And from here arises the most fundamental constants that surround us:

Charge of the electron ... ... $e = 1.602 \times 10^{-19}$ C

Electric Permittivity of vacuum ... ... $\varepsilon_0 = 8.854 \times 10^{-12}$ F / m

Planck constant ... ...$h = 6.626 \times 10^{-34}$ Js

Speed of light ... ...$c = 3 \times 10^{8}$ m / s

All these constants can be practically reduced into one:
Fine Structure constant ... ... $\alpha = 7.297 \times 10^{-3}$
$\approx 1 / 137$
... Which has a direct relationship with all the forces of Nature that surround us ... The classical equation of the electric field is given by the Coulombs law, in which:

$$F_e = K \cdot \frac{Q^2}{r^2}$$

But this equation can also be defined as the following relation:

$$F_e = K \cdot \frac{Q^2}{\lambda_e^2}$$

242

In which it is replaced the value of the distance r by the length of Compton λ_e. The Compton wavelength relates three fundamental constants:

$$\lambda_e = \frac{h}{m_e.c}$$

Replacing in the previous equation, we obtain the following development:

$$F_e = K. \frac{Q^2. m_e^2.c^2}{h^2}$$

$$F_e = \frac{Q^2. m_e^2.c^2}{4\pi. \varepsilon_0. h^2}$$

Where 'm' is the particle which emits and receives the corresponding radiation. In this case, m_e is the mass-energy of the electron and is related with the Electromagnetic Radiation.

In the case of Gravitational Radiation, we could first begin to evaluate our 'gravitational constant' G.

Let us try to determine the origin of this constant…

Chapter XXVIII

DEDUCTION OF THE GRAVITACIONAL CONSTANTE G

"Something it is only impossible until someone doubts
and proves the opposite."
- Albert Einstein -

According to the model presented to our Gravitational Force, we believe that the component of attraction between bodies, the part that attracts and unites the atoms and matter together, is not based in the classical and traditional source of Gravity produced by masses. As so, we assume that this new kind of attraction has no source in masses at all.

As mentioned earlier, this constant of attraction between bodies is not directly related to the quality of mass but is rather related with another characteristic of matter, more precisely with a magnetic component, whose source is in the magnetic spin moments which all particles have.

Despite the strength of this gravitational-magnetic force is clearly small, in the order of magnitude of 10^{-40}, this force of attraction can reach long distances in space and go through all the Universe; since we consider that there is a transfer of the magnetic moment to the macrocosms through the classical way of the Electromagnetic Force. If in Classical Physics the kinetic angular moment can be transferred, than we can also assume that the magnetic moment of a particle can also be transferred.

The clear evidence of the emission of this double radiation to the interstellar space (an electromagnetic force and a gravitational-magnetic force) is always present in the unfolding and decomposition of the Hydrogen spectrum, which always gives us two distinct stripes instead of one single line. Understanding the origin of this radiation and the meaning of these double stripes belongs to the area of Spectroscopy, a science which studies the interaction of electromagnetic radiation with matter, in which, in this case we also have to include the origin of this gravitational-magnetic radiation.

Spectroscopy is nowadays in the great development of current science. From Chemistry to Astrophysics, this new science is revealing new concepts and new possibilities for the interpretation of the Electromagnetic Spectrum; new information previously unknown to us.

As we know, all atoms emit electromagnetic Radiation. The emission of this radiation, of photons, can be produced by various and different subatomic particles such as protons and electrons in motion.

Radiation can have various kind of manifestations and be produced in several ways, such as: Ultraviolet, Visible light, Infrared, Radio waves, etc.. The origin and processing of these different types of radiation is not yet fully understood and the initial explanations and investigations are now beginning to happen, producing major developments in the understanding of the electromagnetic spectrum, showing the reason to be for all this rich interaction between light and matter.

In general, we can say that the emission of electromagnetic waves is related to changes in the energy levels of atoms. For example, the absorption and emission of Visible Light has a direct relationship with transitions between energy levels of the valence electrons (the electrons more distant of the nucleus). The energy of light, and consequently the frequency, and therefore the color, is directly related to the energy difference involved between the two states of the electron transitions in these two energy levels. In the Hydrogen atom this radiation is associated with the Balmer Series.

However, the issue from various sources of radiation is not directly related to the transitions of the energy levels of the electrons! The diversity of this phenomenon is dependent on the variation of the energy of the atom itself. Every time the atom gains or losses energy, radiation is involved. The condition of the frequency of Bohr tells us that:

$$f = (E_i - E_f)/ h \Leftrightarrow \Delta E = h.f$$

It is the transformations and energy transitions of the atom which are on the origin of the emission of different kinds of frequencies, allowing a contribution to the wide range of the entire electromagnetic spectrum.

Several factors may contribute to the variation of the internal energy of the system, thus mean, the atom. Basically, what we assist is that there are changes in the values of the Kinetic Energy and Electrostatic Potential Energy, which can be expressed by several changes, such as: translational motion; rotational motion; vibrational motion; transitions of the electronic levels of the electrons; and changes in the orientation of Spin, nuclear and electronic, of different particles. The phenomena involved in the origin of these different forms of radiation are always distinct.

The contribution to the total energy of the system, of the atom, takes several variables and can be considered as follows:

$$E_{total} = E_{translation} + E_{rotation} + E_{vibration} + E_{electronic\ levels} +$$
$$+ E_{electronic\ spin\ orientation} + E_{nuclear\ spin\ orientation}$$

We have for example, the absorption and emission of Infrared radiation as a result of quantization of the vibrational energy of molecules. Molecules present vibrating movement around their centers of mass. These normal modes of vibration may be present in the direction of the chemical link (distention and elongation) or perpendicular to the chemical connection (bending or deflection angle). As a consequence of this disturbance, of this vibration, the atom is accompanied by a change in its dipolar moment. Only molecules that produces changes in the dipole moment can produce spectrum of IR (Infra-Red).

Other changes in the energy values of atoms can occur and thus produce other forms of radiation. It is known that the emission of Microwave is related to the transition of the rotational energy levels, thus mean, with the rotation of the molecule or atom, and consecutively with the change of the electronic spin orientation.

The emission of Radio waves is related to the transition of the spin energy levels within the core. This radio wave is directly related with the change of the direction of particles spin within the nucleus, such as protons and neutrons (nuclear spin). The way this radio frequency is processed is really interesting ...

Summarizing, any atom or molecule alone has a certain amount of energy associated with the Kinetic Energy and Electrostatic Potential Energy which arises from the state of motion of electrons; and also other smaller quantities of energy associated with the positions and orientation of particles in relation to the centers of mass of the atom or molecule considered. Only certain frequencies, vibrational amplitudes, and certain rates of rotation are allowed for an atom or molecule in particular. Each possible combination of electronic levels, vibrations, rotations, and spin orientation defines a particular level of energy and a very specific frequency, and therefore an emission / absorption of the electromagnetic spectrum. Only certain discrete energy changes are allowed. As provided by the quantum theory, a certain amount of energy is associated with a corresponding emission of a photon radiation.

Most of the absorption lines are associated with orbital transitions (change of electronic distribution) and within this process we can include: the X-ray, Ultraviolet and Visible radiation. Vibrational changes are usually associated with the Infrared. Rotational changes are usually assigned to the Microwave region. Finally, the emission of Radio waves is related with the change of orientation of the nuclear spin. However, the electromagnetic spectrum does not end here, it still continues and shows us another form of radiation...

RADIATION / SPECTRUM	ORIGIN OF THE PHENOMENON	ENERGY BY PHOTON ΔE	FREQUENCY f	WAVE LENGTH λ
Cosmic Ray	Nuclear reaction (production of electron-positron pairs)	> 1,022 MeV	10^{23} Hz Ionizing Radiation	$(10^{-13}$ m$)$
Neutron Radiation	Nuclear reaction (free neutrons penetrating into the nuclei causing radioactivity)		Ionizing Radiation	
Gamma Ray	Nuclear transitions (excited nucleons – reorganization of the nucleus)	10^{-12} J (5 MeV)	10^{21} Hz Ionizing Radiation	Pico- meters $(10^{-12}$ m$)$
Beta Ray	Nuclear Mutation (neutron-proton mutation – emission of an internal electron)	Depending of the radioactive isotope	Ionizing Radiation	
Alpha Ray	Nuclear Fissions (Helium nuclei positively charged)	Depending on radioactive isotope	Ionizing Radiation	

X Ray	Electronic Transitions (electronic layers reorganization -electron transition from a highest to a lowest level)	10^{-15} J	10^{18} Hz Ionizing Radiation	Nano-meters (10^{-9} m)
Ultra violet	Electronic Transitions (Energetic transitions of the valence electrons)	6×10^{-17} J (3,7 eV)	10^{17} Hz	(10^{-8} m)
Visible Light	Electronic Transitions (Electronic transitions of the valence electrons)	10^{-18} J	10^{15} Hz	Micro-meters (10^{-6} m)
Infrared	Vibrational Transitions (alterations of the energy levels by molecular vibrations - variation of the dipolar moment)	10^{-21} J (0,37 eV)	10^{12} Hz	Milli-meters (10^{-3} m)
Micro wave	Rotational Transitions (alterations of the energy levels by molecular rotations – electronic spin variation)	10^{-24} J (0,0037 eV)	10^{9} Hz	Meters (10^{0} m)
Radio Waves	Nuclear Transitions (alteration of the energy levels of the core/atom - nuclear spin variation)	10^{-27} J	10^{6} Hz	Kilo-meters (10^{3} m)
Gravita tional Waves	Nuclear transitions (alteration of the energy levels of the core/atom – nuclear spin alignment with the universal spin and resonance)	10^{-40} J	10^{-6} Hz	Kilo-meters (10^{14} m)

- Electromagnetic spectrum and its origins /
Natural sources of the emission of Radiation -

Within this process the variation of the ranges of energy becomes smaller as the wavelength increases and, consecutively, the energy of the photon irradiated is also lower.

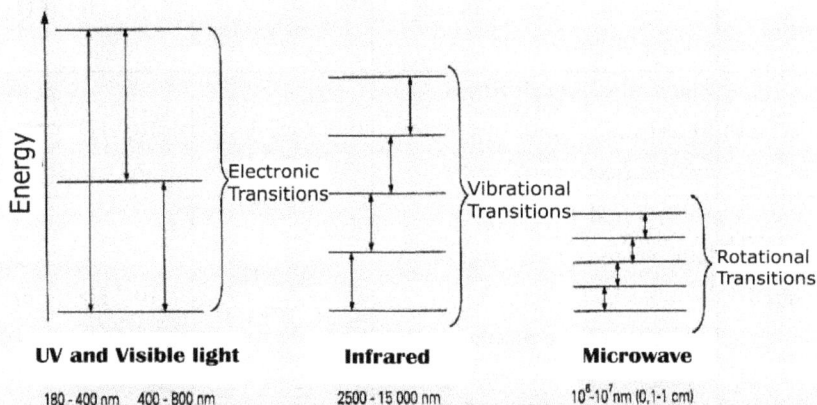

- Variation of the energy levels according to

the different manifestations of radiation -

In certain types of molecules it can occur a simultaneous transition of vibration and rotation. This continuous movement allows the spectrum of absorption show clusters of very close lines, instead of a unique defined transition with a very clear frequency. The existence of these sub-levels of vibration and rotations represent the various possible transitions of energy around a fundamental and center frequency, which is reflected in the absorption spectrum.

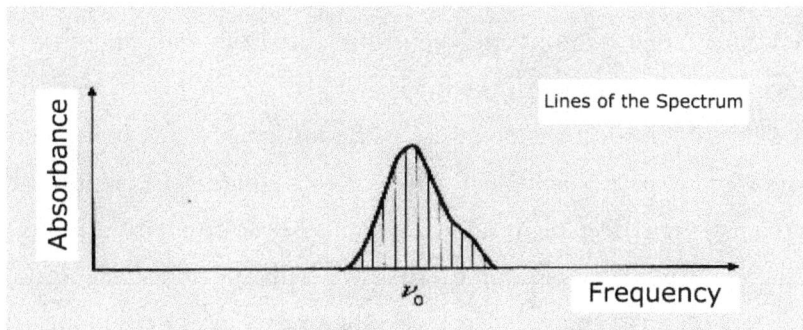

- Relationship of the Frequency with the Spectrum of Absorption -

However, a more careful analysis of the spectrum of Hydrogen atoms shows the existence of two well-defined lines.

In my point view, the reason to be and the explanation of this phenomenon – the double frequency - remains unsatisfactory. The Hydrogen spectrum shows that there are two levels of absorption/emission of two distinct energies, very well defined...

We can say that without rotation, the electrons and protons only have an Electric Force produced by their own electrical charges. But because these particles have a movement of rotation around themselves, a Spin movement, a bit like a pivot that rotates around a vertical axis, these particles acquires a new force, a Magnetic Force, created by a magnetic dipole that installs and involves the particle, also called as a magnetic moment; or intrinsic angular moment; or more simply spin. The electronic spin moment and the orbital angular moment of the electron in motion around its orbit on the atom are combined together to contribute to the total angular moment of the atom.

However, the core of an atom behaves as if it has an independent nuclear magnetic moment. Since each particle of the nucleus produces nuclear magnetic interactions with the environment around them. We

251

cannot forget that protons and neutrons also have spin and that they interact to a contribution for the nuclear spin.

Another interesting characteristic is the intrinsic magnetic moment of the subatomic particles, which when placed under the action of an external magnetic field B_0, it only has two possible orientations: plus and minus: ½, which correspond to the two only possible values of the magnetic potential energy. The two alignments of nuclear spins are, therefore, manifestations of different energies, according to their orientations. These spin directions are classified as follows:

Spin -½ aligned against the field
(anti-parallel);

Spin +½ aligned with the field
(parallel).

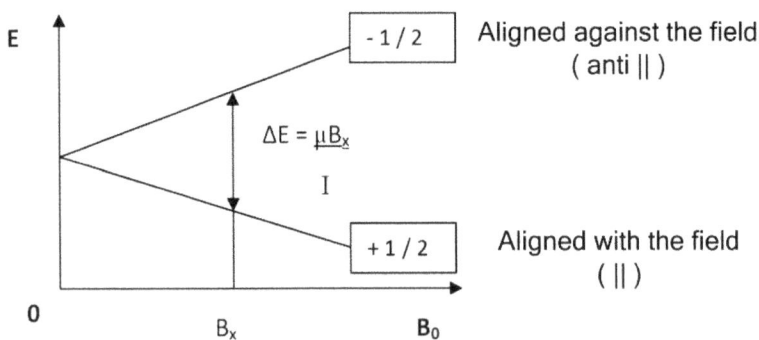

- Orientation of the magnetic moment of Spin μ_s -

A chemical species that emits radio frequency, thus mean Radio Waves, is the atom of Hydrogen ^1H, the simplest atom of the Periodic Table and, coincidentally, the most abundant type of atom in the Universe.

Each atom of Hydrogen has one proton and one electron in its constitution, and both particles revolve around their axes and have spin. The direction of the spins of these two particles can only have two possible orientations: the proton's spin is parallel to the spin of the electron and both operate in the same direction, or each particle spins is in opposite directions and the spin is anti-parallel.

The state of lowest energy of the Hydrogen occurs when the spin of the nucleus (proton) is opposite to the spin of the electron. However, the atom of ^1H can receive external energy and with it produce a parallel spin alignment of the proton and the electron. Since the Hydrogen atoms are continuous moving from the parallel configuration (maximum energy) to the anti-parallel configuration (lowest energy level), the excess energy acquired is released in the form of radiation: Radio Waves.

The characteristic wavelength of this emission is about 21 cm, and with a frequency equal to 1.4 Gigahertz. There is a predominance and a preference for the parallel || orientation, but the difference is very small, only an excess of 10 in a total of 10^6 nuclei is presented in the state of spin of higher energy.

The atom of Hydrogen vacillates and constantly oscillates between these two states and the strength of energy it gets comes from the application of its own external magnetic field, produced by the orbital motion of the electron. The absorption of energy occurs when the magnetic field of the nucleus B' (proton) is aligned with the external magnetic field B_0 (of the electron). This nucleus aligned with the field absorbs extra energy, changing its spin orientation into the opposite direction.

- Parallel orientation of the spin magnetic field B' (or H') -

Usually a macroscopic magnetic field is defined by the quantity vector B (Tesla), also known as magnetic induction. However, when referring to magnetic fields at a microscopic level we can use another scale which relates the intensity of the magnetic field, defined as H (Ampere / Meter).

All nuclei have a characteristic spin (designated by a 'I' letter), depending on the number of protons and neutrons that enter into their constitution. Thus, some nuclei have fractional spins: I = 1/2, 3/2, 5/2 ... others have integer spins: I = 1, 2, 3 ... and some do not have any spin at all: I = 0, since the contribution of the nuclei, with a pair number of protons and neutrons, produces coupled spins with a total spin zero.

Nuclei containing an odd number of protons and neutrons (an unpaired number of nucleons) have a quantized spin, as so, they have a magnetic moment. Examples of chemical nucleus with a magnetic moment associated are: ^1H, ^{13}C, ^{19}F, ^{31}P, etc. To highlight again the Hydrogen! The most abundant chemical element in the Universe contributes to a significant nuclear magnetic moment ...

When we refer to molecules we must considered that, besides the interaction with B_0, the nuclear spins can also feel the presence of other spins in the molecule. If a molecule has got several atoms in its constitution, this condition leads that the atmosphere of a single magnetic spin is also depending of the orientation of other magnetic spin moments neighboring the molecule. Spins do interfere with each other.

The magnetic characters of these nuclei and of these atoms are naturally weak and we could think that this low magnetism would not have any magnetic influence with other particles more distant in space. However, if the magnetic fields of each nuclear spin is presented in a regular and orderly position in a certain moment in time, if these domains stay all aligned during a short moment, then, as it happens at the Magnetization phenomenon we may say that these single magnetic spins spread through the Universe may feel other spins from far away atoms quite distant from each other. And, this would be the process which would lead to a standardized and universal spin magnetic alignment, consisting of the same direction and orientation. In this state of Resonance the magnetic flow becomes higher and, consecutively, this represents a gain of extra energy for the atom which, afterwards, it will immediately release this extra energy into space in the form of electromagnetic emission or … Gravitational Radiation!

There is a moment in time when it occurs a synchronized alignment between all the magnetic moments; as so, we can imagine the existence of a specific coupling where all spins are in phase … the presence of a Universal Spin!

These magnetic interactions are a quite complex. When considering a single atom of Hydrogen, what we have, in this case, is a single electron, unpaired, associated with an atomic nucleus which has a considerable magnetic moment. Therefore, this electron will feel not only the external magnetic environment produced by other atoms in its neighborhood and from himself but also it will feel the magnetic field emitted by the core, thus mean the nuclear proton.

We know that the component of the magnetic orbital moment of the electron (translation movement) contributes to the total magnetic moment of the atom and it is related with the classical Electromagnetic Force. However, we cannot ignore that there is another subatomic magnetic component. The existence of an individual magnetic component (the spin magnetic moment due to the movement of rotation) which all particles have, contributes to the magnetic field of each particle, creating a new magnetic field, smaller and independent of the Electromagnetic Force. This new field - when considering a nuclear particle of the atom and not the atom itself - is related to the classical Gravitational Force, let's see how:

In a general standard we can consider that the magnetization of an atom is equal to the magnetic moment per unit of volume. The resulting magnetization is directly related to the total magnetic field inside the atom (of the constituent particles of the atom), and it is also depending on the external magnetic field which is being applied (of the orbital motion of the electron). So, the total magnetic field (B) produced by an atom depends on the contributions of the orbital magnetic moment of the electron (μ_L) and also of the magnetic spin moments (μ_S) of its constituent particles.

In a similar analogy, when considering a region of space where there is a magnetic field B_0 produced by a wire conductor crossed by an electrical current; and if we fill this region with a magnetic substance, the total magnetic field B produced in this region will be according to the following expression:

$$B = B_0 + B_m$$

Where B_0 is the field introduced and B_m is the field caused by magnetization of the substance, and this is directly dependent on the magnetization vector M:

$$B_m = \mu_0 M$$

We believe that our total field B depends on the contribution of two different magnetic fields: B_0 and B_m.

We can draw this analogy to our atoms; considering that the total magnetic moment is dependent of two different magnetic moments: the angular or orbital moment μ_L and of the spin magnetic moment μ_S.

In our case concerns us, particularly, the field produced and induced by the magnetization process, thus mean, B_m.

Given that all the vast region of the Universe is mostly filled with atoms of Hydrogen; and according to data provided by the Critical Density, we can consider that the average density of the Universe is about 6 atoms of Hydrogen per unit volume. As we know, these atoms of Hydrogen have in its constitution one electron and one proton. Both particles contribute with their own magnetic moments: the spin magnetic moment S for the proton and the electron, and the angular magnetic moment L for the orbital electron. All these moments contribute to the resulting magnetic moment of the atom. The maximum magnetization occurs according to the vector sum of all these magnitudes. Imagining that in a certain moment in time all these vectors have the same orientation, the resulting magnetic moment μ_r will be given by the following equation:

$$\mu_r = \sqrt{(\;\mu_{Se}^2 + \mu_{Sp}^2 + \mu_{Le}^2)}$$

$$\mu_r = 1{,}313 \times 10^{-23}\ \text{J/T}$$

Where we have:
Spin magnetic moments of the electron and proton:
$$\mu_S = -\,g_s m_s \mu_B:$$

$$\mu_{Se} = 9{,}285 \times 10^{-24}\ \text{J/T}$$
$$\mu_{Sp} = 1{,}410 \times 10^{-26}\ \text{J/T}$$

And the orbital angular magnetic moment of the electron, also known as Bohr's magneton $\mu_B = -e/(2m_l)\ L$:

$$\mu_B = \mu_{Le} = 9{,}285 \times 10^{-24}\ \text{J/T}$$

The maximum magnetization or the saturation magnetization is obtained through the following expression:

$$M = n.\mu_r$$

Where n defines the number of atoms per unit of volume and μ_r the resulting magnetic moment.

Considering that our sample corresponds to the entire Universe, we will consider as a reference the critical density, which corresponds to 6 atoms of ^{1}H by m^3.

$$M = 6 \times (1{,}313 \times 10^{-23})$$

$$M = 7{,}878 \times 10^{-23} \text{ A/m}$$

Returning to our expression to obtain the magnetic field B_m induced by the magnetization process and considering μ_0 as our magnetic constant, is that:

$$B_m = \mu_0 M$$

$$B_m = (4\pi \times 10^{-7}) . (7{,}878 \times 10^{-23})$$

$$B_m = 9{,}9 \times 10^{-29} \text{ T}$$

According to the classic formula, a Magnetic Force is obtained through the next equation:

$$F_m = Q.v.B \text{ sen}\theta$$

This force is maximum for θ = 90 °, thus mean sen90 ° = 1, and simplifying:

$$F_m = Q.v.B$$

Making the substitution of the values in this equality, and considering the average speed of action as v = c/137 = 2,2 x 10^6 m/s, it comes that:

$$F_m = (1,6 \times 10^{-19}).(2,2 \times 10^6).(9,9 \times 10^{-29})$$

$$F_m = 3,48 \times 10^{-41} \text{ N}$$

Most curiously, the order of magnitude of this magnetic force that arises is framed in the same order of magnitude of the Gravitational Force!

Very interestingly …

Considering the possibility that the gravitational force (F_g) is a magnetic force (F_m) and matching both equations, is that:

$$F_g = F_m$$

$$G.m^2/r^2 = Q.v.B$$

$$G = \frac{Q.v.B.r^2}{m^2}$$

$$G = F_m \cdot r^2 \cdot m^{-2} \qquad N. \, m^2. \, Kg^{-2}$$

Thus, one can evaluate another evidence in our 'Gravitational Constant', which are the units that appears in the relationship between the gravitational force and magnetic force, because they do match perfectly!

In this way, the Gravitational constant G arises as a Magnetic Constant!!

Considering 'r' as the atomic radius of Hydrogen (r_H = 25 pm); and the total mass 'm' as the contribution of the mass of the electron and the mass of the proton: m = $((1,672 \times 10^{-27})+(9,109 \times 10^{-31}))$ ⇔ m = $(1,673 \times 10^{-27})$ kg; and replacing the values in the equality, we have:

$$G = \frac{(3,48 \times 10^{-41}).(2,5 \times 10^{-11})^2}{(1.673 \times 10^{-27})^2}$$

$$G = \frac{(2,175 \times 10^{-62})}{(2.799 \times 10^{-54})}$$

$$G = 7,77 \times 10^{-9} \qquad N.m^2.Kg^{-2}$$

This would be the G reference value surrounding an atom of Hydrogen and diluted in an interstellar space filled mostly by vacuum and in accordance with the data supplied to the critical density we can considerer an average of 6 Hydrogen atoms per m^3.

The magnetic moment of the Hydrogen is indeed immense. Since the proton has a very significant magnetic moment and the highest of all the fundamental particles. Thus, in interstellar clouds composed of gas, dust and other materials, it is among these chemicals substances that begins the first gravitational attractions and the formation of the first aggregations of matter. Mother Nature gives priority to the formation of these clouds of Hydrogen, made up mostly of atomic and molecular Hydrogen (H_2).

With the passage of time, the clustering of this Hydrogen mass is gradually compressed by action of its own Gravity Force. Being constantly and successively compressed, the density and temperature slowly increases, gradually warming this giant molecular of mass and forcing it to revolve faster and faster on its own. When the gas heats up enough and the core temperature reaches 10^7 Kelvin, begins the nuclear reaction that causes the fusion of Hydrogen into Helium. At this time, nuclear fusion occurs and the protostellar cloud becomes into a new born star.

The interstellar environment and the formation of these nebulae is of extreme importance in the evolution of the cosmos. These are prolific areas par excellence. It is in this interstellar space which is born all the new generations of stars which exist in our Universe.

However, the measurement of the gravitational constant is dependent on several factors. As we have seen previously, when we referred to the experience of Cavendish, we saw that the influence of temperature has a decisive role in the outcome of the G constant. As a final result and as a consequence, is that this gravitational constant must reach very big values in stars. Similarly, the rotation speed of the star is also very important when measuring this constant. Therefore, the final conclusion is that, this constant must therefore reach the highest values in neutron stars and pulsars.

Returning to Planet Earth and when we refer to planets in general we should consider other chemical elements more complex and more dense which participate in its formation, such as iron (^{26}Fe) for example, constituent of the Earth's core (which contributes, at least, with one electron with an unpaired spin); among many other chemical elements with specific characteristics and very particular proprieties. Considering that these elements have different atomic radius, whose value is also influenced by external conditions such as pressure and temperature, but in general we can consider that the average radius of an atom is in the order of magnitude of 10^{-12} m.

Among the atomic radius of Iron; the average Bohr radius; and the Compton wavelength, we can begin to consider, for example:

$$r = 2,31 \times 10^{-12} \text{ m}$$

With this small change in the atomic radius it gives us a substantial difference in the gravitational value, the G constant:

$$G = \frac{F_m . r^2}{m^2}$$

$$G = \frac{(3,48 \times 10^{-41}).(2,31 \times 10^{-12})^2}{(1.673 \times 10^{-27})^2}$$

$$G = 6,6 \times 10^{-11} \quad N.m^2.Kg^{-2}$$

And so we get the average value of G at the surface the earth!

In fact, the accurate measurement of this constant must require a highly complex process ... since this constant is always inconstant in each location.

But the fact is that this constant can be influenced by two decisive factors:

- By the variation of the atomic radius (r);

- And by the rate of change/intensity of the external magnetic field (v)

(the speed variation of the magnetic field is influenced by the speed of rotation of the star and temperature).

Therefore, the universal constant G is a magnetic constant and a variable constant!!

While everyone thinks that the origin of Gravity hides no secret ... When everyone 'knows' that the source of Gravity comes from the quality of mass; this simple feature so much 'clear' for us in Nature ... It's at the obvious that no one notices and that no one doubts, and it is exactly in that propriety, in this particular characteristic, that Nature surprises us completely!

After all, it was nothing like the way we had imagine, it was nothing like the way we were thinking...

THE TRAVEL IN TIME

" *I have no special talent.*

I'm just extremely curious."

- Albert Einstein -

EPILOGUE

"As we understand things better,
everything becomes more simple. "
- Edward Teller -

Although we assume and constantly recognize that what we have in our hands is an incomplete and unsatisfactory Law of Gravity, it will now be very difficult to insist going at the same direction knowing that we have the knowledge of a new theoretical model that fits with the experience; which is consistent with facts; that presents us solutions to resolve imperfections at the old Theory of Gravity. We can no longer still continue to work with classical gravitational theory and insist on finding solutions that it cannot give us, since the basic fundaments on which this theory is based are incorrect.

It may be that this New Theory of Gravity passes by overlooked and not given the correct attention of the circle of professors and studious interested in this thematic. Suggesting this revolutionary change, which is in agreement with facts, should be enough to convince the traditional science. However, this way of thinking has already been studied and patiently explained by the physicist Ed May:

Normally we assume that Science is a rational process, but it is not quite so. Usually what happens when a new theory is presented is that we will, inevitably and naively, make the mistake of the 'rational man'. When we are confronted with evidence and proofs which contradict our most fundamental beliefs, instead of this new evidence lead us to a new belief and to a new understanding of a subject, in general, what happens is exactly the opposite.

Challenging an ancient belief leads to reaffirm more strongly our previous beliefs! This reluctance to change comes from a phenomenon also quite simple, which is: Far too often History repeats itself. Unfortunately, prejudice, self-indulgence, skepticism or more simply ignorance, are the most common responses when our most fundamental beliefs are threatened, and these confrontations can be perpetuated for centuries! Maybe one day this New Theory of Gravity will be seen in a most common and natural way, having nothing of extraordinary or highly magical. As it is the fact today that we consider the Sun at the center of the solar system and not the Earth! So, perhaps the 'revolutionary' adjective may not be the most appropriate.

To justify the results obtained previously we will develop the physical process between Coupling and Angular Momentum.

Considering only the information given by the electronic distribution of an atom, we know that this is not sufficient to fully describe the energy state of the atom, since it experimentally appears that several orbitals with the same electronic configuration can have different energy levels.

The variation of this energy is related to the orbital angular momentum L and also with the spin angular momentum S, which is combined to determine the total angular momentum resulting for the atom J.

In order to calculate the total angular momentum of an atom it's only necessary to consider the angular momentum of the valence electrons, because, the total angular momentum of each complete layer is equal to zero. These angular momentum have variable directions and different orientations over time, and they are combined according to specific rules of Quantum Mechanics, so that, the final angular momentum J can be determined.

A vector model image will serve us to define the composition and space-time evolution of all these angular momentum:

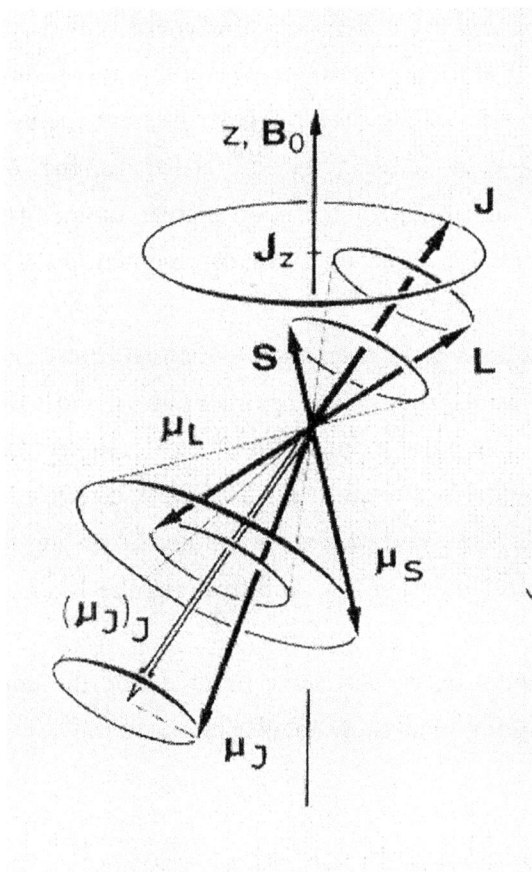

- Relation between angular momentum L, S and J
  and magnetic moments μ_L μ_S μ_J -

The search for a complete Formalism of Unification still continues...

"All that I only know is that I know nothing."

- Socrates -

"I think, therefore I exist."

- Descartes -

But do I think right?!

- C. P. Fournier -

EVERYTHING IN THIS BOOK MAY BE WRONG

According to the theory of Eduardo Punset, this neuroscientist describes the process of creativity in a particular and interesting way: In general, people follow other people just because there are many people going in that direction. However, a creative person decides to be independent. For example, if a creative person sees that everyone is moving the same direction, then he decides to move exactly the opposite way, not accepting the majority as the correct direction.

The creator thinks: 'I have my own idea and maybe my idea is better'. A creative person thinks in a different way from the typical or common people, although this attitude sometimes entails its consequences.

You must know that when we have a creative idea, a new idea, others will not accept it easily. In a way, it is necessary and even essential to apply some effort to try to demonstrate that our idea is valid and that it may bring some advantage. In practice, the creator is simply trying to convince others that his idea may be, perhaps, a better idea. So, the author of a new idea develops all the necessary skills so that his new idea is not immediately rejected and neglected. Considering that, at least, a new idea deserves to be part of the discussion. Although there only is one substantial difference, which is that the creator deeply believes in what he has created...

DEDICATION AND ACKNOWLEDGMENTS

I would like to thank to my first reader, António Saraiva, author of numerous articles published in the Scientific Electronic Journal: 'The General Science Journal'; Who has shown the kindness of generously donated his time to read the manuscript, the complete literary book, contributing with a very positive opinion. Mentioning his own words: "It's a noble lesson of Physics, explained in a very simple way (…) which ideas deserve to be published.".

I also want to make a special thank to the editorial director of the IST PRESS, my second reader, Prof. Dr. Joaquim Moura Ramos, teacher at the 'Superior Technical Institute' in Lisbon, who has also contributed with a very positive assessment, expressing a scientific literary criticism also positive, appreciating the work invested and the new ideas developed. His words were simple: "Delightful to be read.".

A sweet thank to Paul Fournier and to João Figueirinha for their valuable notes and comments, and for having found the 'Travel in Time' fascinating, extraordinary, and absolutely 'interesting'!

Finally, I could not forget to express my special attention to Walter Babin in Canada, editor of the online journal 'The General Science Journal' - a journal with worldwide recognition and an opportunity for any writer to publish their articles, including this one -. Quoting his own words: "You have a delightful and informative way of writing and it is a pleasure to read it.".

Again, my sincere thanks to all these people.

Finally, this book would make no sense if I did not mention one last dedication:

In memory of all physicists;

And to my wonderful Universe.

FINAL NOTE

I Would like to mention that any errors that remain in this work, the literary or scientific concepts, are the sole responsibility of the author. Well, actually, I never had any class of Cosmology, Quantum Physics or Relativity Theory!

"I burn inside as a sacred fire when I think of them and I feel I'm never tire of repeating them. To conclude this letter, let me enjoy the pleasure of transcribing them again: 'I do not care if my work will be read now or by posterity. I can wait a century for readers (...) I shell win!'
I have stolen the secret of the golden stars.
I satisfy myself with my own sacred fury. "
- Johannes Kepler -.

*

THE END

Vol I – The Travel in Time

C. P. FOURNIER

www.ingramcontent.com/pod-product-compliance
Lightning Source LLC
Chambersburg PA
CBHW060336200326
41519CB00011BA/1953